Flux Cored Arc Welding Handbook

Third Edition

William H. Minnick

Professor Emeritus
Palomar College
San Marcos, California

Publisher
The Goodheart-Willcox Company, Inc.
Tinley Park, Illinois
www.g-w.com

Library of Congress Catalog Card Number 200804917

ISBN 978-1-60525-077-9

2 3 4 5 6 7 8 9 – 09 – 18 17 16 15

Library of Congress Cataloging-in-Publication Data

Minnick, William H.
    Flux cored arc welding handbook / William H. Minnic. — 3rd ed.
        p. cm.
    Includes index.
    ISBN 978-1-60525-077-9
1. Electric welding. I. Title. II. Title: FCAW handbook

TK4660.M526 2009
671.5'212—dc22   20088040917

Introduction

Flux Cored Arc Welding Handbook provides a simple but complete introduction to flux cored arc welding. It covers principles, equipment, techniques, modes of operation, and safety in a straightforward manner.

Flux cored arc welding is an important commercial and industrial welding process. Improvements in the process, as well as refinements in equipment over the years, have established its position as a major industrial welding system.

Careful study of this book and practice with welding procedures will provide the knowledge and skill you need to secure employment as a welder. The chapters will lead you step-by-step through the principles and practice of flux cored arc welding. Topics include:

- Basic operation of each component of the flux cored arc welding system.
- Safe practices directed specifically toward working with electricity, shielding gases, and other hazards common to welding.
- Various types of welds and weld joints.
- Welding techniques and procedures for various metals.
- Weld defects and how to avoid them.

For a beginning student of welding, the book provides the background needed for a successful welding career. Advanced students will value this book for its detailed coverage on welding various metals using the different modes common to the industry.

Every weld you make bears your personal trademark and represents your skill as a welder. Take pride in your work, and make each weld as if your career depended on it.

William H. Minnick

About the Author

Realizing the need for a specialized type of welding text for instructors and students, **William H. Minnick** has drawn upon his many years of experience as a welder, welding engineer, and community college instructor to develop this text for training future welders.

The author's career in industry includes welding jet engines, missiles, pressure vessels, and atomic reactors. He developed the welding procedure for, and welded, the first titanium pressure vessel for the Atlas missile program. He has written many technical papers, including articles on research and development for welding exotic materials and modification of existing welding processes for automatic and robotic applications.

Mr. Minnick has developed welding certificate and degree programs, and taught all phases of welding and metallurgy in community colleges for more than twenty years. In addition to this **Flux Cored Arc Welding Handbook,** he is the author of **Gas Tungsten Arc Welding Handbook** and **Gas Metal Arc Welding Handbook.**

*C*ontents

Acknowledgments

The author gratefully acknowledges the assistance of the following companies who contributed suggestions, ideas, photographs, and information to this textbook.

Air Reduction Co.
Airco
Arcal Chemical, Inc.
CK Systematics, Inc.
Distribution Designs, Inc.
Eutectic Corp.
G.A.L. Gage Co.
G.S. Parsons Co.
Golden Empire Corp.
Hobart Brothers Co.
Inco Alloys International, Inc.
Jetline Engineering, Inc.
L-Tec Welding and Cutting Systems
Lincoln Electric Co.

Linde Co.
M&K Products, Inc.
Magnaflux Corp.
Miller Electric Mfg. Co.
Nederman, Inc.
Tescom Corp.
Thermco Instrument Co.
Tweco Products, Inc.
Union Carbide
Veriflow Co.
Victor Equipment Co.
Weld World Co.
Western Enterprises Co.

The author is deeply grateful for the many tables, figures, and information from the American Welding Society, the computer drawings from my son Steven A. Minnick, and the photographic printing from my son William R. Minnick.

William H. Minnick

CHAPTER 1

Flux Cored Arc Welding Process

Objectives

After studying this chapter, you will be able to:
- Define flux cored arc welding.
- Identify the two flux cored arc welding processes.
- Distinguish between semiautomatic and automatic modes of operation.
- Name the equipment and supplies used in flux cored arc welding.
- Recall the advantages, disadvantages, and applications of flux cored arc welding.

Important Terms

constant current
constant voltage
direct current electrode negative (DCEN)
direct current electrode positive (DCEP)
gas-shielded arc process
open arc process
oxidation
power supply
shielding gas
wire feeder

Definition

As defined by the American Welding Society, *flux cored arc welding* (FCAW) is an electric arc welding process that fuses metallic parts by heating them with an arc between a continuously fed, consumable, flux cored electrode and the work. Primary shielding for the molten weld metal is obtained from the fluxing ingredients within the metal tubular electrode that are deposited as slag on the top of the weld. The flux cored electrode may be supplemented by an externally applied shielding gas. **Figure 1-1** shows the operating end of the welding gun, weld joint metal, and weld slag coating.

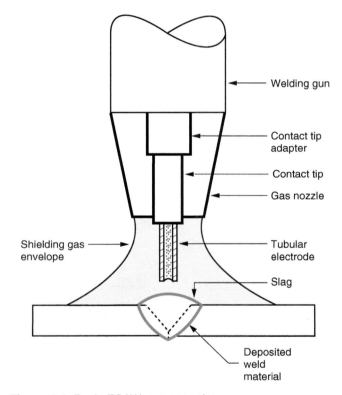

Figure 1-1. Basic FCAW gun operation.

Process Types

Two types of processes are used in FCAW for the deposition of molten metal. In the **open arc process,** (FCAW-ss), all of the fluxing ingredients required for the proper transfer of the filler material and shielding of the molten pool are included in the core material.

In the **gas-shielded arc process,** (FCAW-g), carbon dioxide is used alone, or in combination with argon in a specified mixture of the two gases, to shield the molten stream and weld metal from the outside atmosphere.

Modes of Operation

A FCAW process may be carried out in one of two modes. In a *semiautomatic operation,* the weld is made using process equipment and a manual welding gun. See **Figure 1-2.**

In an *automatic operation,* the weld is made using specialized equipment and controls that are monitored by the operator. Automatic welding is often used for large weldments to increase production and maintain greater quality control. See **Figure 1-3.**

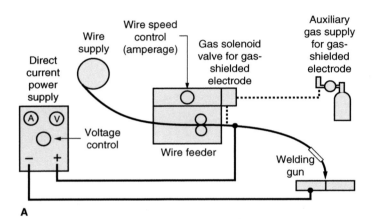

A

B

Figure 1-2. Semiautomatic welding operation. A—Equipment for a FCAW semiautomatic system employing the gas-shielded arc process. B—A welder uses a semiautomatic system.

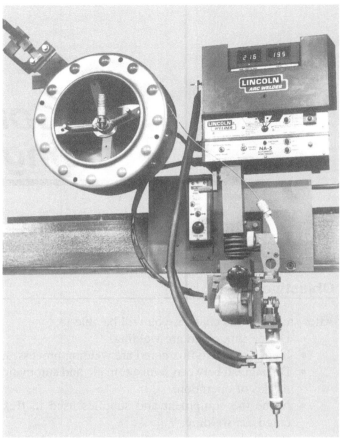

A

B

Figure 1-3. Automatic welding operation. A—FCAW automatic system. (Lincoln Electric Co.) B—An automatic system is used to weld a steel vessel. (Jetline Engineering, Inc.)

Welding Equipment and Systems

A *power supply* provides the appropriate type and amount of welding current needed to melt the base material and filler wire (electrode). See **Figure 1-4.** The flux cored arc welding process uses two types of current (amperage):

- *Direct current electrode negative (DCEN),* also known as *direct current straight polarity (DCSP).*
- *Direct current electrode positive (DCEP),* also known as *direct current reverse polarity (DCRP).*

A *constant voltage* (CV) power supply unit is generally used for FCAW. A *constant current* (CC) unit may be used with a different type of wire feeder.

The *wire feeder* is an electromechanical device that feeds the required amount of filler material at a fixed rate of speed throughout the welding operation. See **Figure 1-5.** The filler material is supplied in coils or wound on spools that are mounted on the wire feeder.

When using the gas-shielded process, gas-regulation equipment is required. An adequate supply of *shielding gas* is needed to permit a continuous flow of gas during the welding operation. See **Figure 1-6.**

Guns are held by the welder or mounted on the machine during the welding operation. They contain components (parts) that transfer electrical current to the wire. Triggers or switches start and stop the operation. See **Figure 1-7.** If a gas nozzle is used, it is attached to the end of the gun to direct the shielding gas around the wire and the molten pool. This protects the wire and the

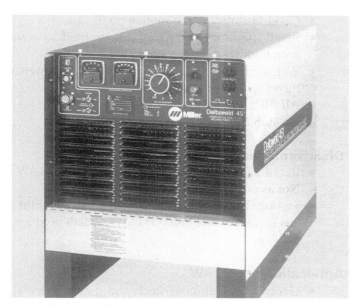

Figure 1-4. This constant voltage power supply may be used for semiautomatic or automatic welding operations. (Miller Electric Mfg. Co., Inc.)

Figure 1-5. This wire feeder may be used for semiautomatic or automatic welding operations. (Miller Electric Mfg. Co., Inc.)

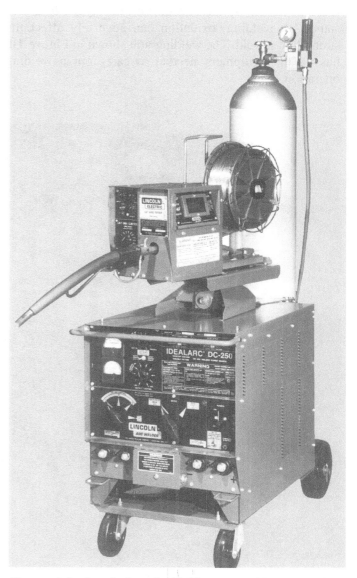

Figure 1-6. A cylinder of shielding gas with an attached regulator is mounted with a power supply and wire feeder on a cart for portability. (Lincoln Electric Co.)

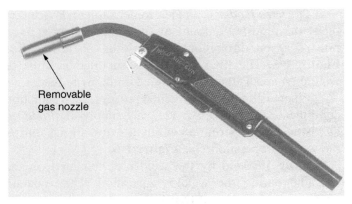

Figure 1-7. This gun has a nozzle for use during the gas-shielded process. The nozzle can be removed for the open arc process. (Tweco Products, Inc.)

molten metal from oxidation. ***Oxidation*** is a chemical reaction caused by the introduction of oxygen to a substance. In welding, oxidation can adversely affect the strength of a weld. The welding unit shown in **Figure 1-8** has all the equipment needed to carry out a welding operation.

Figure 1-8. A stationary, fully assembled welding unit. (Hobart Brothers Co.)

Advantages and Disadvantages

FCAW is a major welding process utilized by many industries to fabricate and maintain numerous types of welded components. The process has some advantages and disadvantages compared to other welding processes.

Advantages of FCAW
- Both semiautomatic and automatic applications.
- Very good metal deposition rates.
- Very good metallurgical values.
- Open arc mode can be used outdoors in a slight wind.
- Electrode flux cannot be broken off, as occurs in shielded metal arc welding (SMAW).
- Appeals to operators, so training time is reduced.
- Good visibility of the molten pool.
- Small diameter electrodes can be used to weld with high productivity on thin material.
- Welds can be made in all positions with higher travel speeds.
- All filler material is consumed.
- No stub loss, as occurs in SMAW.

Disadvantages of FCAW
- Requires more complex equipment than SMAW.
- Not as portable as SMAW.
- Produces slag, which must be removed, unlike gas metal arc welding (GMAW), which produces no slag.

Applications of FCAW
- General fabrication of steels and stainless steels.
- Structural steel fabrication of buildings, bridges, ships, and containers.
- Pressure vessels for various gases and liquids.
- Construction equipment such as bulldozers, earthmovers, backhoes, and tractors.
- Railroad equipment, switches, and tracks.
- Farm equipment repair, modifications, and hardfacing.
- Automobile and truck frames.
- Shelves, bins, and storage compartments.
- Surfacing components for various types of wear.
- Repair of home garden tools, rakes, and shovels.
- Maintenance and repair of castings, conveyors, handling equipment; hardfacing of cutting tools.

Review Questions

Please do not write in this text. Write your answers on a separate sheet of paper.

1. The _____ is the organization that defines the standard welding processes used in the welding industry.

2. How are metallic parts fused in the flux cored arc welding process?
3. Where are the fluxing ingredients for the primary shielding of the FCAW process located?
4. What are the two processes used for deposition of molten material?
5. What gases may be used in the FCAW-g process?
6. When the welder deposits the weld metal using a hand-held gun, the operation is called _____ welding.
7. When a welding operator uses specialized equipment and monitors the process, the operation is called _____.
8. What are the two types of supply current used in FCAW?
9. The two types of power supplies used in the FCAW process are constant _____ and constant _____.
10. An electromechanical device that feeds the required amount of filler material at a fixed rate of speed is called a(n) _____.
11. Insufficient gas coverage will cause the weld wire and molten metal to _____, diminishing the metal's strength.
12. Which is an application of FCAW?
    A. Hardfacing.
    B. Repairing castings.
    C. Wear surfacing.
    D. All of the above.

The welder is using a wire feeder and power source for flux cored arc welding.
(Miller Electric Mfg. Co.)

CHAPTER 2 *FCAW Operation and Safety*

Objectives

After studying this chapter, you will be able to:
- Describe how an arc is established.
- Distinguish between the open arc and gas-shielded arc processes of metal deposition.
- Identify the types of power supplies used in FCAW.
- Recognize wire feeders, gas cylinders, and welding guns.
- Cite safety precautions regarding electrical current, shielding gas, the welding environment, other concerns.

Important Terms

arc gap
auxiliary shielding gas
contact tip
Dewar cylinders
regulator/flowmeter
solenoids
stickout
turbulence
welding guns
wire feeders

Establishing an Arc

In flux cored arc welding, a consumable tubular electrode is fed through a welding gun. Heat from an electrical current melts the electrode and base metal to make a fusion weld.

The direct current flows from the power supply, through the welding or ground cable to the welding gun, through the contact tip, to the consumable tubular electrode. An arc is established when the welding electrode contacts the workpiece and completes an electrical circuit. Arc length is controlled by the conditions preset on the power supply, or in some cases, by the speed of electrode travel.

Metal Deposition

The FCAW process deposits high-quality weld metal because the proper fluxing agents have been added to the electrode core. In some cases, additional shielding of the arc and molten metal may be required to ensure the transfer of all core materials.

Weld metal deposited in the open arc process utilizes the flux within the electrode core for shielding the arc and molten metal. A thin layer of slag is then deposited on the surface, **Figure 2-1.**

Weld metal deposited in the gas-shielded arc process utilizes the flux within the electrode core, plus an *auxiliary shielding gas,* to protect the arc stream and molten pool from atmospheric oxygen and nitrogen. Carbon dioxide or an argon-carbon dioxide mixture is used as the shield. A thin layer of slag is also deposited

Figure 2-1. In the open arc process, slag is deposited on the work surface. This process is often used for outdoor welding where wind drafts or gusts are a problem.

on the surface of the completed weld, **Figure 2-2.** For proper shielding, the weld should be protected from wind and drafts.

Stickout is the length of electrode that extends beyond the contact tip during the welding operation. Stickout distance is controlled by the movement of the gun to and from the work surface in both semiautomatic and automatic operations. Manufacturers of filler material establish the correct stickout for each type of electrode.

On a nozzle with a recessed contact tip, the distance of the stickout viewed by the welder is referred to as the *visible stickout.* Manufacturers of filler material use the term *electrode stickout (ESO).*

A *normal stickout* distance is used when making general welds (fillet, butt, edge, corner, lap) with the standard weld requirements as to shape, size, and contour. See **Figure 2-3.**

A *long stickout* distance is used when making high-deposition welds (fillets, grooves, buildup, surfacing) in the flat or horizontal position. Specially designed nozzles are needed to protect the electrode from exposure to the atmosphere during the operation, **Figure 2-4.** The long stickout utilizes the electrode extension to

Figure 2-2. The gas nozzle should be large enough to allow the gas shield to cover the molten weld metal.

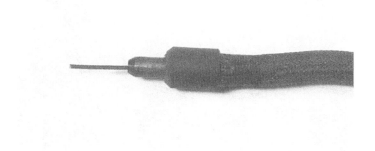

Figure 2-3. Open arc gun with a normal stickout. Changing the required stickout will affect the amount of welding heat at the weld pool.

Figure 2-4. On a long stickout, a nozzle must be used to shield the electrode from the atmosphere and prevent oxidation at high temperatures.

preheat the wire for more rapid melting, which increases metal deposition.

Weld Setup

Proper weld deposition depends on test-weld procedure results. Test welds are made with specific base materials, type and diameter of electrode, and joint design. Welding is done in the required position to determine the equipment setup and to establish the welding parameters for the actual weld.

Parameters are the range of operating values and include:
- Amount of welding current (wire speed).
- Amount of arc voltage (arc gap).
- Type of shielding gas, flow rate, and nozzle size, if used.
- Stickout.
- Travel speed.

During the procedure test, many other *variables* are tested and developed. These operations include:
- Preweld cleaning.
- Preheat, interpass, and postheat temperatures (if required).
- Electrode positions and techniques.
- Bead sequence.
- Welding progression.

Power Supplies

The basic power supply used for FCAW is a constant voltage (CV) power supply. It is also called a constant potential (CP) machine. See **Figure 2-5.** The machine supplies varying amounts of amperage to maintain the preset welding voltage (*arc gap*). Voltage is adjusted during the welding operation to change the arc gap dimension as needed.

In some cases, a constant current (CC), or variable voltage (VV), machine may be used to supply power. However, these machines require additional components in the system for the process to operate properly.

Figure 2-5. A constant voltage power supply uses a voltage control to adjust arc length. (Lincoln Electric Co.)

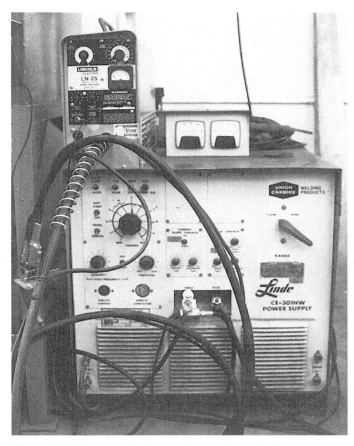

Figure 2-7. On a constant current power supply, the amperage is set on the machine. The arc voltage is set on the special voltage-sensing wire feeder.
(L-Tec Welding and Cutting Systems, Lincoln Electric Co.)

Power supplies may also include:
- CV combined with a wire feeder in a single unit, **Figure 2-6.**
- CC combined with a specially designed feeder, **Figure 2-7.**
- Portable CC and CV units for field welding where utility power is not available, **Figure 2-8.** The units are gasoline- or diesel-powered, air- or liquid-cooled, and use a wire feeder similar to the type shown in Figure 2-7.
- Inverter-type CV machines. They are small compared to standard-type power supplies and use a standard or portable feeder, **Figure 2-9.**

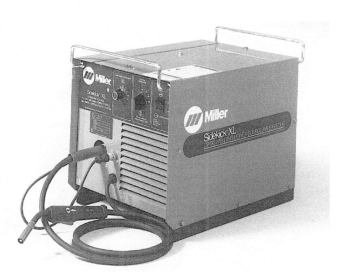

Figure 2-6. A power supply and wire feeder combination unit that operates on 115 V ac. Voltage and feed controls are located on the front panel. (Miller Electric Mfg. Co.)

Figure 2-8. Portable welding supplies and ac generators are available for use with gasoline or diesel fuels.
(Lincoln Electric Co.)

Figure 2-9. Inverter technology reduces the size of the power supply while maintaining capacity and duty cycle. (PowCon, Inc.)

Wire Feeders

Wire feeders feed continuous filler metal from coils and spools through cables and conduits to the gun, and finally to the welding arc, **Figure 2-10.** An adjustable control on the welding gun regulates the feed rate of the wire, which must be constant at any speed, **Figure 2-11.** A separate wire feeder is available for use with a constant potential power supply, **Figure 2-12.** A *voltage-sensing* wire feeder may be used with a constant current or constant voltage-type power supply, **Figure 2-13.** Portable units are also available, **Figure 2-14.**

Gas Supply and Regulation

Shielding gases are sometimes used to prevent contamination of the welding electrode and weld. Argon

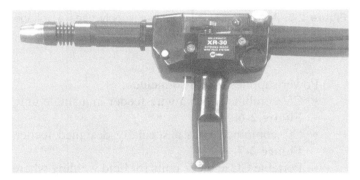

Figure 2-11. A welding gun contains an electric motor and drive rollers that pull the filler metal from the main drive system located in the feeder. (Miller Electric Mfg. Co.)

Figure 2-10. The wire feeder and supply spool are enclosed in the main welding unit, eliminating the need for another piece of equipment. (Miller Electric Mfg. Co.)

Figure 2-12. This wire feeder is used only with a constant potential power supply. (Lincoln Electric Co.)

Figure 2-14. This portable wire feeder is designed for use with direct current (dc) power only. However, changes in polarity can be made by a switch without changing the leads from the power supply. (Lincoln Electric Co.)

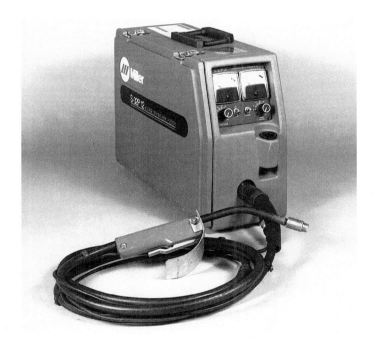

Figure 2-13. This special wire feeder is designed to operate on the welding voltage from the constant current power supply and does not require 115 V ac. (Miller Electric Mfg. Co.)

and carbon dioxide are the most common gases used. They are supplied by the manufacturer as a single or a mixed gas, in a liquid or gaseous state, in pressurized cylinders up to 3000 psi (20 700 kPa). *Regulators* are used to reduce the pressure to a workable amount. A *flowmeter* is used to regulate the amount of flow to the weld area. A *regulator/flowmeter* is commonly used for high-pressure

cylinders, **Figure 2-15.** The flow of gas is controlled in cubic feet per hour (cfh) or liters per minute (lpm).

Sufficient gas must be delivered to the welding area. Too little gas flow allows air to enter the weld zone and contaminate the weld. Too much gas flow causes *turbulence* (great changes in speed and direction of flow) in the gas envelope, allowing air to enter the weld zone and contaminate the weld. Whenever atmospheric gases enter the weld zone, the quality of the weld diminishes.

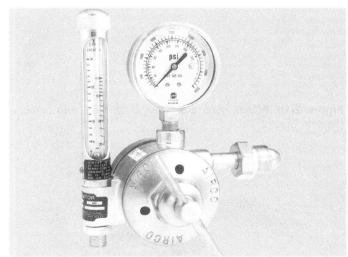

Figure 2-15. A single-cylinder regulator/flowmeter has a gauge showing cylinder pressure. Gas flow to the gun is regulated by turning the adjustment knob and reading the ball position in the vertical tube. (Air Reduction Co.)

Electrically operated valves (*solenoids*) are included in the system to start and stop the flow of shielding gas from the gas supply to the welding gun. The valves are actuated by closing the switch on the manual gun or the switch in the arc starting circuit. These valves are usually located in the wire feeder unit.

Welding Guns

The FCAW process uses *welding guns* designed for specific applications. Guns are rated for their current-carrying capacity and the length of time they are used at the given amperage. Some do not use any cooling at all, some use gas cooling, and others are water-cooled for heavy-duty service. The gun is constructed from materials that protect the user from electrical shock and provide a means to conduct electrical current to the consumable welding electrode as it passes through the *contact tip* at the end of the welding gun. See **Figure 2-16.**

Gas nozzles of various designs and diameters attach to the gun to direct the shielding gas around the electrode and the molten metal. Welding electrodes that require gas shielding are used with guns that have gas nozzles attached, **Figure 2-17.** In a water-cooled, gas-shielded gun, an electrical motor and drive rolls within the gun body pull the filler material from the supply spool, **Figure 2-18.**

Other welding guns may have suction devices attached to the nozzle to remove smoke and vapors from the weld area, **Figure 2-19.** When welding is done by a

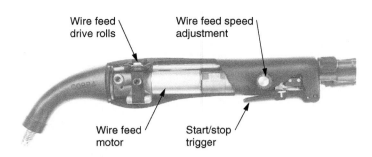

Figure 2-18. Water hoses attached to the rear of the body give this water-cooled, gas-shielded gun a very high duty cycle. (M&K Products, Inc.)

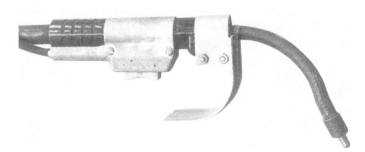

A

B

Figure 2-19. A smoke-removal vacuum system. A—A special nozzle attached to the gun removes smoke vapors from the welding operation. B—This vacuum unit is used in conjunction with the special smoke removal gun nozzle. (Lincoln Electric Co.)

Figure 2-16. FCAW guns are heavy-duty, safe, and reliable. (Lincoln Electric Co.)

Figure 2-17. A typical gas-shielded gun. (CK Systematics, Inc.)

robot, the gun may be directly mounted onto the wire feeder. A wire straightener mounted on top of the feeder is used to remove some cast from the electrode before it enters the feed rolls, **Figure 2-20.**

Safety Precautions

FCAW is a skill that can be performed safely with minimum risk if the welder uses common sense and follows safety rules. Make it a habit to practice safety. Start by regularly checking the equipment and the welding environment. Safety rules fall into four categories: electrical current, shielding gas, the welding environment, and special hazards.

Electrical Current Safety

Primary current to the electrically powered welding machine is usually 115 V ac to 440 V ac or more. This amount of voltage can cause extreme shock to the body and even death. The following safety rules must be strictly followed:

- Never install fuses of higher amperage than specified on the data label or the operation manual.
- Always use a ground wire from the machine frame to a suitable solid ground.
- Install electrical components in compliance with all electrical codes, rules, and regulations.

- Be certain that all electrical connections are tight.
- Never open a welding machine cabinet when the machine is operating, unless you are a competent electronic technician and fully trained in machine installation and operation.
- Always lock primary voltage switches open, and remove fuses when working on electrical components inside the welding machine. *Then, put the key to the lock in your pocket!*
- Welding current supplied by constant current (CC) and constant potential (CP) power supplies has a maximum open circuit voltage of approximately 80 V. At this low voltage, the possibility of lethal shock is small. However, it can still produce a shock strong enough to cause health problems. To reduce the chances of this happening:
  - Keep the welding power supply dry.
  - Keep the power cable, ground cable, and gun dry.
  - Make sure the ground clamp is securely attached to the power supply and workpiece.

Shielding Gas Safety

The gases used in FCAW are produced and distributed to the user in either a liquid or gaseous state. All gas storage vessels must be approved and stamped by the Department of Transportation (DOT). Previously this was done by the Interstate Commerce Commission (ICC). See **Figure 2-21.**

Some gases used in FCAW are inert and colorless, so special precautions must be taken when handling

Figure 2-20. Guns used for robotic welding are often water-cooled to allow high-amperage duty cycles. (Linde Co.)

Figure 2-21. This cylinder is made and tested to comply with Department of Transportation (DOT) specifications. The letter "T" designates the size of the cylinder (330 cu. ft. capacity). The name is that of the owner.

them. All FCAW gases are *nontoxic*; however, they can cause asphyxiation (suffocation) in a confined area that lacks sufficient ventilation. Any atmosphere that does not contain at least 18% oxygen can cause dizziness, unconsciousness, and even death.

Shielding gases cannot be detected by the human senses and can be inhaled like air. Never enter any tank, pit, or vessel where gases may be present until the area is purged (cleaned) with air and checked for oxygen content.

High-pressure cylinders contain gases under extreme pressures of 2000 psi to 4000 psi (13 800 kPa to 27 600 kPa) and must be handled with extreme care. The following rules for the storage and handling of cylinders must be strictly followed:

Storage:
- Store all cylinders in the vertical position.
- Always use a cylinder cart to move a cylinder. Secure the cylinder with a safety chain, cable, or strap. *Never roll a cylinder!* See **Figure 2-22.**
- Keep the protective cylinder cap in place until the cylinder is ready for use.

Handling:
- Use the proper equipment for the type of gas used.
- Know the contents of the cylinder. Cylinders should be labeled, **Figure 2-23.**
- Before attaching the regulator, check the outlet threads, and "crack" (slightly open and close) the cylinder valve to clean the valve opening. See **Figure 2-24.**
- Properly attach the regulator, and securely mount the flowmeter in the vertical position, **Figure 2-25.**

- When opening a cylinder valve, the regulator adjusting screw should always be free of pressure on the regulator diaphragm.

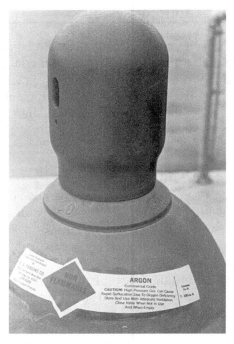

Figure 2-23. The label indicates that argon gas is stored in this cylinder. Note the safety cap is in place when the cylinder is not in use.

Figure 2-22. Always transport a cylinder with a cylinder cart, and secure the cylinder with a safety chain.

Figure 2-24. Cracking a cylinder before attaching a regulator blows out dirt in the valve opening.

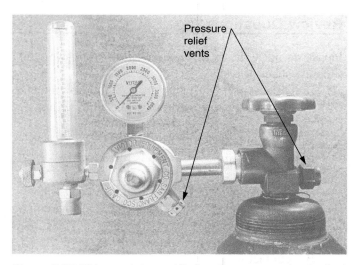

Figure 2-25. This regulator and flowmeter are properly attached to the cylinder. The pressure-relief vents on the regulator and cylinder serve to release excess pressure in the cylinder. Never tamper with these devices. (Victor Equipment Co.)

Figure 2-26. A Dewar cylinder contains very cold liquid gases. A heat exchanger is needed to convert the liquid to a gas. This unit has an auxiliary heat exchanger mounted on top of the cylinder to increase the amount of liquid conversion to gas.

- Stand to one side of the valve front, and slowly open the cylinder valve. Never stand in front of the gauges when opening a cylinder.
- When a cylinder is not in use, the cylinder valve should be closed, and the diaphragm screw should be released.
- When a cylinder is empty, close the valve, replace the safety cap, and mark the letters "MT" (empty) with chalk on the upper part of the cylinder.
- Never tamper with a leaky valve; return the cylinder to the supplier for a replacement.

Liquefied gas cylinders, commonly called *Dewar cylinders,* are essentially vacuum bottles. For ease of storage and handling, the gas is reduced to a liquid at the supplier's plant. Conversion of the liquid to a gas for welding is done by heat exchangers within the cylinder or as a part of the gas-delivery system in the welding facility, **Figure 2-26.** The following safety rules apply to the Dewar system:

- Always keep cylinders in the vertical position.
- Always use a cylinder cart to move a cylinder. Dewar cylinders are extremely heavy and difficult to handle.
- Always use the proper equipment when installing or connecting cylinders.
- Do not interchange equipment components.
- Liquid gases are extremely cold and will cause severe frostbite to exposed eyes or skin. Always wear gloves and safety glasses. Do not touch any frosted surface with your bare hands.

Welding Environment Safety

A clean, well-organized, and well-maintained shop is a source of pride as well as a necessity for a safe environment. The following rules for the welding environment must be followed:

- Keep the welding area clean, well-ventilated, and free of combustibles.
- Repair or replace worn or frayed ground and power cables.
- Make sure the part to be welded is securely grounded.
- Make sure welding helmets have no light leaks.
- Use a proper filter plate lens to protect your eyes from arc radiation. (See Chart R-1 in the Reference Section.)
- Wear safety glasses when grinding or power brushing.
- Wear tinted safety glasses when others are tackwelding or welding nearby.
- Use safety screens or shields to protect your work area.
- Wear proper clothing. Your entire body should be covered to protect you from arc radiation.
- When welding on cadmium-coated steels, copper, or beryllium copper, use special ventilation to remove fumes and vapors from the work area, **Figure 2-27.** Smoke removal guns can also be used.
- Do not weld near trichloroethylene vapor degreasers. The arc changes the vapor to a gas. A sweet taste in your mouth indicates that this gas is being formed.

Figure 2-27. This portable smoke collector removes smoke from the welding operation. The tube and collector can be placed in any position without auxiliary support. (Nederman, Inc.)

Special Hazards

Be aware of these special hazards and the precautions you must take to ensure shop and personal safety:

- Fires may be started by the welder in a number of ways, including igniting combustible materials, misuse of fuel gases, electrical short circuits, and improper ground connections. Be careful not to start a fire. *If you do start a fire, make sure it is completely extinguished before you leave the area.* Know where the fire extinguishers are located.
- Never weld on a container that previously held a fuel until you are sure the container has been purged with an inert gas and tested for fume content.
- Never enter a vessel or confined space that has been purged with an inert gas until the space is checked with an oxygen analyzer to determine that sufficient oxygen is present to support life.
- Never use oxygen in place of compressed air. Oxygen supports combustion and will cause a fire to burn violently.
- Power brushes are very dangerous because they expel broken pieces of wire. Always wear safety glasses or safety shields when using power brushes.
- Be alert to the clamping operation when working with mechanical, hydraulic, or air clamps on tools, jigs, and fixtures. Serious injury may result if parts of the body are exposed to the clamp action.

Review Questions

Please do not write in this text. Write your answers on a separate sheet of paper.

1. In the FCAW process, the arc is started by the electrode touching the workpiece and completing a(n) _____.
2. What is the purpose of an auxiliary shield gas?
3. In the FCAW process, argon is mixed with _____ as a shielding gas.
4. What FCAW process is used outdoors and why?
5. The distance of the wire extension from the contact tip is called the _____.
   A. arc gap
   B. contact tip
   C. stickout
   D. wire feeder
6. What does *visible stickout* refer to?
7. How is correct stickout length determined?
8. High deposition welds are done using a(n) _____ stickout and gas nozzle to protect the electrode from oxidation.
   A. visible
   B. electrical
   C. normal
   D. long
9. Amperage, voltage, travel speed, and shielding gas flow rate are examples of welding _____.
10. *Arc gap* is determined by the _____.
    A. welding voltage
    B. variable voltage
    C. constant current
    D. None of the above.
11. Both constant _____ and constant _____ power supplies are used for the FCAW process.
12. The wire feeder used with a constant current-type power supply is called a(n) _____ unit.
    A. inverter
    B. voltage-sensing
    C. portable
    D. None of the above.
13. _____ are used to regulate the pressure and volume of shielding gas to the welding gun.
    A. Regulators
    B. Flowmeters
    C. Regulators/flowmeters
    D. None of the above.
14. What will happen if too much gas is admitted into the gun nozzle?
15. List two safety rules from each safety category discussed in the chapter (electrical current, shielding gas, the welding environment, and special hazards) for a total of eight safety rules.

3 Equipment Setup and Control

Objectives

After studying this chapter, you will be able to:
- Identify power supply controls and their functions.
- Describe proper power supply installation and maintenance.
- Distinguish three types of electrode feeders.
- Identify feeder controls and their functions.
- Explain the functions of cables and guns.

Important Terms

arc voltage
cast
contact tips
guns
open circuit voltage
power supplies
pull-type feeders
push-type feeders
push-pull-type feeders
slope

Power Supplies

Power supplies are machines that produce current for melting the electrode at a low voltage. The equipment must be able to control the welding operation in several areas:
- Input voltage (primary voltage).
- Open circuit voltage.
- Output ratings and performance.
- Duty cycle.

The National Electrical Manufacturers Association (NEMA) has established specification *EW-1 Electric Arc Welding Power Sources* for control of these areas.

Power Supply Specifications

Each type of power supply is designed for a specific purpose, with limitations established to ensure proper operation. Specifications for machines using utility power fall into the following categories:
- *Primary power type, voltages, and cycles.* Includes alternating current, single- or three-phase power at 110 V, 208 V, 230 V, and 460 V and 60 Hertz (Hz) cycles.
- *Primary power fusing.* Fuse sizes are specified for individual machines. Limits must not be exceeded.
- *Rated welding amperes.* Current amounts specified by the manufacturer should not be exceeded, since the cooling system is not capable of carrying away excess heat.
- *Duty cycle.* All welding power supplies are designed to operate for a specific time period at a specific load. Design considerations include:
  - Size of internal wiring.
  - Type of internal components.
  - Insulation of internal components.
  - Amount of cooling required.

The NEMA specification establishes that every 10% of duty cycle represents one minute of operation in a 10-minute period. See **Figure 3-1.** A 150 ampere (A) machine, for example, would rate a 30% duty cycle for three minutes. Before welding is resumed, the machine would have to idle seven minutes to allow internal components to cool.

Duty cycles range from 20% to 100%. Electric arc welding machines are classified as follows:
- Class I—60%, 80%, or 100% duty cycle
- Class II—30%, 40%, or 50% duty cycle
- Class III—20% duty cycle

Machines not made to NEMA specifications have duty cycles specified by the manufacturers. Never exceed an equipment manufacturer's duty cycle requirements.

Duty Cycle	Number of Minutes Machine May be Operated at Rated Load in a 10-minute Period
100%	. Full time
60%	6
50%	5
40%	4
30%	3
20%	2

Figure 3-1. Power supply duty cycles limit the number of minutes the unit may be operated at the rated load.

Direct-Current, Constant-Voltage Power Supply Controls

The type and number of controls vary with the power supply. They may range from a tap connection and a simple rheostat to numerous controls necessary for high-quality welds, **Figure 3-2.** Typical power supply controls include:

- *Open circuit voltage (OCV) range switch.* **Open circuit voltage** relates to the output values during the welding operation and is established by the power supply manufacturer. The range of voltages may be controlled by adjusting taps, levers, or switches. The basic response output of a power supply can be computed on the machine volt-ampere curve.

When setting up a machine with a voltmeter on the panel, follow these steps to establish the proper OCV range for the arc voltage to be used:

1. Turn the power supply on.
2. Release idler roller pressure to prevent feeding.
3. Place the voltage range switch in the desired location. (Check the manual OCV and the arc voltage range to select the range of values.)
4. Hold the gun away from the ground or workpiece, and energize the contactor switch on the gun. (When you depress the switch, the contactor allows electrical current to flow to the electrode tip and the welding electrode. An arc will form if the electrode contacts the ground.)
5. Observe the voltmeter, and adjust the fine-tuning voltage control to the desired OCV. The arc voltage will be 2 to 3 volts lower for each 100 amperes. See **Figure 3-3.**
6. Release the trigger or switch, and reset the tension on the idler roller.

- *Arc voltage.* The electrical measurement of the arc gap between the end of the electrode and the workpiece is called the **arc voltage.** Adjust the fine-tuning control (the same control used to set the OCV) to the desired voltage. This control may be adjusted during welding.

- *Slope.* In FCAW, **slope** refers to the slant of the volt-ampere curve and the operating characteristics of the power supply under load. In many machines, the slant of the volt-ampere curve is automatically set as the OCV is changed. In other machines, the curve can be adjusted for various welding conditions. FCAW operates best with the power supply set in the flat mode, **Figure 3-4.**

Figure 3-2. This machine is rated 300 A dc and has adjustable voltage, inductance, and slope. (Miller Electric Mfg. Co.)

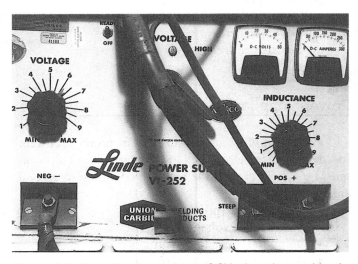

Figure 3-3. The voltmeter registers OCV when the machine is on and the gun switch is open, but welding is not taking place. The voltmeter registers arc voltage during the welding operation.

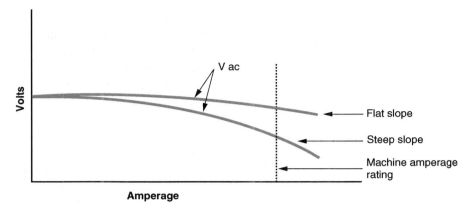

Figure 3-4. The flat OCV range allows greater voltage/amperage variance. The flatter the slope, the more rapid the amperage change.

- *Inductance.* Some power supplies may have an inductance or pinch-effect control on the main panel. However, this control has no effect on the power supply output in the FCAW operation. It is used for gas metal arc welding with the short-arc mode.

Remember: The function of the power supply is to maintain a preset voltage condition. The feeder produces a continuous electrode feed and, when required, the machine adjusts with more or less welding current to maintain the preset arc voltage length. Thus, the machine is called a *constant potential* or *constant voltage power supply* and produces only direct current.

Constant-Current Power Supply Controls

A constant-current power supply is often used for FCAW. A CC machine requires an OCV slope set as flat as possible. The only way to control slope is to place the amperage range switch (if available) in the highest position. The main amperage rheostat is set for the amount of welding current desired to burn off the electrode fed into the molten pool. See **Figure 3-5.**

Power Supply Installation

A power supply should be installed in an area free of dust, dirt, fumes, and moisture. Machine heat must be able to escape. Both dirt and improper cooling can cause a power supply to overheat and the internal components to be ruined. Electronic components that absorb moisture can fail. Make sure the free flow of air into or out of the machine is not blocked by objects. Most machine manufacturers specify the required space for air circulation around a power supply.

Power supplies use various input voltages and 60 Hz cycles. Machines using other than 60 Hz are specially made. The required fuse size for the incoming power is shown on the data label, **Figure 3-6.** The fuse panels should always be close to the power supply so the main power can be disconnected in an emergency, **Figure 3-7.** Machines that do not match the utility power voltage may be used with a step-up or step-down transformer, **Figure 3-8.**

Power Supply Maintenance

Insulated transformers, solid state components, and sound designs have extended the life of the modern

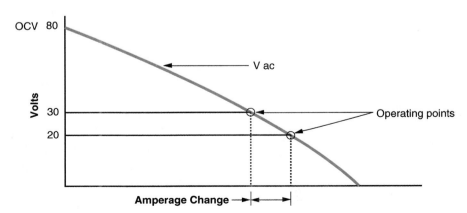

Figure 3-5. With an OCV of 80 V, the highest possible amperage setting must be used to obtain a flat slope value. Set the amperage range switch to maximum, and adjust the machine rheostat for the desired amperage.

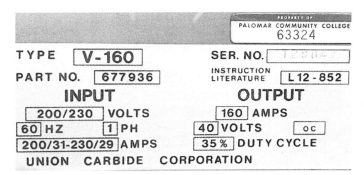

Figure 3-6. The data label on the power supply indicates requirements for incoming power and rated output power.

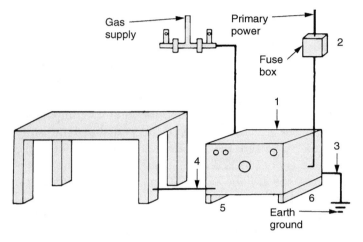

1. Locate power supply away from walls for proper airflow.
2. Locate primary fuse box near power supply.
3. Connect power supply frame to earth ground.
4. Connect work ground with 2/0 min-4/0 max cable.
5. Keep power supply dry.
6. Keep area near power supply clean.

Figure 3-7. A typical power supply installation includes a fuse box mounted near the power supply. An earth ground from the power supply protects the welder from primary line high voltage.

Figure 3-8. Transformers change incoming primary power to match the requirements of the power supply.

welding power supply. Given reasonable care and routine maintenance, a welding power supply will operate satisfactorily for many hours before repairs are needed. When performing maintenance or repair of a power supply, follow these guidelines, along with the manufacturer's instructions:

- Turn off the machine, and disconnect the circuit at the fuse box before beginning work.
- Clean or blow out the unit periodically. Use only dry, filtered, compressed air, nitrogen gas, or an electrical (nonconducting) cleaner. Always wear a face shield when using compressed air or gas.
- Check all terminals for loose connections.
- Lubricate fan and motor bearing, if required.
- Check mechanical arms and switches for freedom of movement. Lightly grease mechanical connections.
- Clean terminal blocks and tap connectors with a wire brush or abrasive pad. If corroded, they will restrict the flow of electrical current. Make sure they are tight.
- Check motor generator brushes, and replace them when worn beyond the manufacturer's tolerances. Worn brushes wear armatures, which then must be machined for proper operation.
- Lubricate, inspect, and adjust portable gasoline and diesel power supplies. Improper maintenance will reduce their capacity and operation.
- Use only authorized replacement parts.

Feeders (Stationary Models)

Most wire feeders operated by industry use either 24 V or 115 V ac power obtained from an outlet on the power supply. If this connection is used, the feeder automatically turns on or off when the power supply is started or stopped.

A feeder used on a push-pull system or gun uses a 24 V dc motor. The motor powers drive rollers installed in hand-held guns for precision drive of the electrode. The low-voltage design ensures the safety of the welder.

Types of Feeders

FCAW uses three basic types of wire feeders:

- *Push-type feeders* push the electrode from the feeder through the cable to the gun, **Figure 3-9.**
- *Pull-type feeders* use driver rollers to pull the electrode from a spool to the contact tip. They are used when operating with a spool on a gun, **Figure 3-10.**
- *Push-pull-type feeders* push the electrode through the cable to a separate feeder mounted in the gun handle. Such a system is used for driving electrodes long distances at a constant rate of

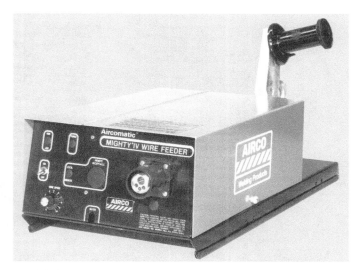

Figure 3-9. A push-type feeder pushes the electrode to the welding gun. (Airco)

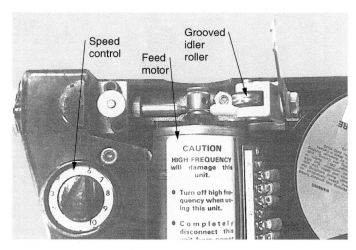

Figure 3-10. In a pull-type feeder, the drive rolls are mounted in the gun head to feed the electrode from the supply spool. (Miller Electric Mfg. Co.)

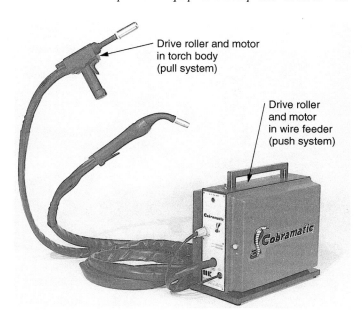

Figure 3-11. Push-pull systems work very well, even when the electrode is a considerable distance from the feeder. (M&K Products, Inc.)

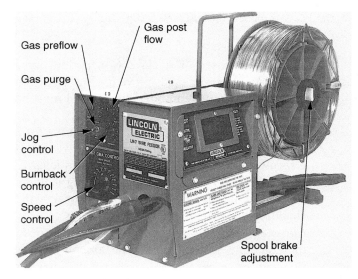

Figure 3-12. Manufacturers place various controls on the feeders depending on the use of the machine. (Lincoln Electric Co.)

speed, **Figure 3-11.** The drive unit, located in the main feeder, is called a *slave unit.* It pushes the electrode to the gun or feed rolls. Then, the feeder pulls the electrode to the work area.

Feeder Controls

The number and type of controls included in the wire feeder vary depending on its use and the amount of feed desired. In many cases, the feeder is designed to operate with both the GMAW and FCAW processes, **Figure 3-12.** The major controls of a dual-process feeder include:

- Off/on switch.
- Feed potentiometer (speed control).
- Spool brake control (stops the spool at end of welding).

The major controls of a welding machine/feeder combination, **Figure 3-13,** include:

- *Mode control.* Used to place the machine in modes such as spot, stitch, and seam welding.
- *Trigger lock-in control.* Allows the weld operator to depress the start switch to begin the weld cycle and to continue the weld without depressing the trigger. To end the weld sequence, the operator depresses the switch and releases the trigger.
- *Burnback timer (antistick control).* Sets the length of time the arc current remains on after the stop

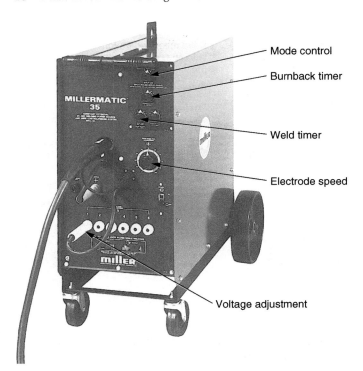

Figure 3-13. Many auxiliary controls may be placed on a power supply and feeder. Some, but not all, of the controls are needed for the FCAW process. (Miller Electric Mfg. Co.)

A

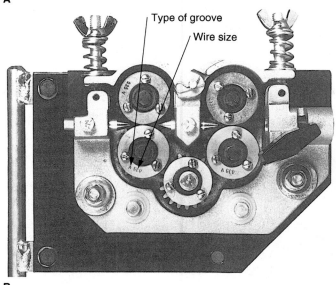

B

Figure 3-14. Feeder drive roll systems. A—Two-roll drive system. B—The electrode size is stamped on the outside rollers of this four-roll drive system. (Miller Electric Mfg. Co.)

switch is released. The current-on period burns back the electrode and prevents the tip from sticking in the molten pool. This control must be used when spot welding.

- *Spot weld timer.* Controls the length of time for a spot weld operation. A *stitch weld timer* controls a specific time period for a specific weld length.

Controls used for FCAW with gas shielding include:

- *Prepurge timer control.* Sets the length of time the shielding gas flows before welding starts.
- *Purge control.* Allows the weld operator to open the shielding gas solenoid. The amount of gas flow can be set on the flowmeter, or the operator can purge the gun gas lines prior to welding.
- *Postflow timer control.* Allows the weld operator to establish a timed period for the gas to flow after the welding is completed.
- *Electrode inching control (electrode jog control).* The electrode can be moved from the feeder to the gun or to the stickout distance. (**Note:** In some cases the jog and run speeds are the same, and electrode speed may be measured using this control.)

Feeder Drive Roll System

Either a two- or four-roll system drives the electrode from the spool, **Figure 3-14.** Four-roll feed systems are used where precision drive is required. Always use drive rollers suited to the type and diameter of the filler

electrode. Do not adjust feeder drive rolls too tightly. An electrode straightener is used to increase or decrease the amount of coil *cast* (curvature). Incorrect cast of the filler material may cause arc outages due to lack of current flow between the contact tip and the electrode.

Drive rolls may be knurled, grooved, or cogged in "V" or "U" shapes, **Figure 3-15.** The "V" grooved roll is

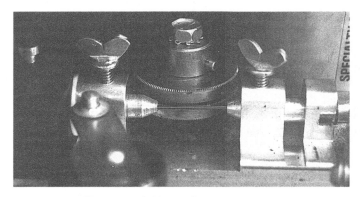

Figure 3-15. The type of driver roller used depends on the type and diameter of filler material. (Miller Electric Mfg. Co.)

used for hard-casing-type (also called Innershield) electrodes. The "U" grooved roll is used for soft-casing-type (also called Outershield) electrodes. The serrated drive rolls are not used in the FCAW process.

Solid rolls or split types with variable thickness spacers between the rolls are used for different diameters of electrode. Use the correct diameter roller for the type and size of electrode; otherwise, the material may split, or "birdnesting" will occur at the drive roller, **Figure 3-16.** Adjust the idler roller so slippage does not occur.

Electrode Wiper

A wiper is an accessory attached to the inlet side of the drive roll system, **Figure 3-17.** It usually consists of a piece of felt material secured around the electrode with a clothespin. Adding a little oil to the felt lubricates the electrode and helps extend the life of the guides and the cable liners. Do not over-oil.

Figure 3-16. "Birdnesting" is caused by improper operation of the feed roller system. Check the guidelines, rollers, and tension to correct the problem. (Miller Electric Mfg. Co.)

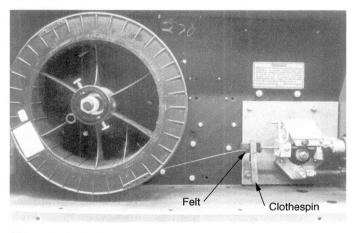

Felt   Clothespin

Figure 3-17. A wiper assembly can be made from a clothespin and piece of felt. Commercial fluid or light grade oil may be used as a lubricant.

Feeders (Portable Semiautomatic)

A portable semiautomatic unit is used for welding at a distance from the basic power supply and may be used in either a CC or CV mode, **Figure 3-18.** When used in the CV mode, the unit is set up as a standard feeder. When used in the CC mode, the unit uses a voltage-sensing motor to drive the electrode at the proper speed. Before welding begins, the speed and amperage relationship must be established from a setup chart, **Figure 3-19.** (A complete setup chart is shown in Chapter 7.) The correct polarity is established on the feeder, and a voltage-sensing clip is installed on the workpiece.

The voltage adjustment will not operate if the unit has a voltage-adjustment dial on the face of the feeder and the power supply is CC. The arc voltage must be set using a setup chart. **Caution:** Unless a special circuit has been installed on the drive unit, the electrode will be electrically "hot" (charged) when connected to an operating power supply.

Feeder Maintenance

As with all electrical devices, feeders must be kept dry. Malfunctions should be checked by a skilled

Figure 3-18. This portable feeder may be used in the CC or CV mode. A self-shielded or gas-shielded electrode may also be used since a gas-flow system is installed in the unit. (Lincoln Electric Co.)

Figure 3-19. A setup chart is installed within the feed compartment. When using a CC machine, the amount of current desired must be set on the power supply. (Lincoln Electric Co.)

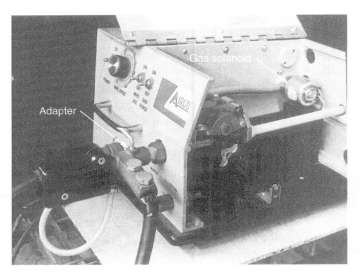

Figure 3-20. Different types of guns can be connected to a feeder by using an adapter. (Air Reduction Co.)

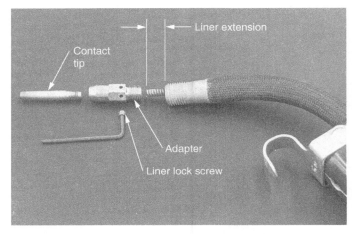

Figure 3-21. Gun liners must fit into the adapter within a specific dimension. An incorrect fit will cause the electrode to feed improperly. (Tweco Products, Inc.)

electrician. Any printed circuit boards should be replaced with factory-authorized parts obtained from a welding supplier. Electrical motor brushes should be checked and replaced periodically.

Inspect the guides and rollers often, and replace them when worn beyond specifications. If the electrode is not feeding properly, the problem probably exists in the guides and rollers. To determine if the feeder is at fault, disconnect the gun cable, and run the electrode through the feeder only. If the feeder runs smoothly, connect the gun and cable to the feeder. Operate the feeder. If the problem continues, the gun and cable assembly is at fault. Be certain the cable assembly is free of kinks or sharp bends and the contact tip is in good condition.

Gun Cables

Gun cables are used to carry electrical current, gas, welding electrode, cooling water, electrical circuit wire (from the start switch), and vacuum hoses (for smoke removal). With all of these operations to perform, cables must be cared for properly. Keep gun cables as straight as possible. Protect them from falling objects and wheels running over them. Cables come in various lengths, and some can be connected to make longer assemblies.

Adapters are available for attaching different guns and feeders, **Figure 3-20.** Gun manufacturers make adapters for almost every type of feeder.

A liner is installed in the cable to protect it from wear. Specific liners are made for use with certain sizes of electrodes. Always follow the manufacturer's installation instructions. Liners are not interchangeable, **Figure 3-21.**

Guns

Guns are designed to carry electrical current to the contact tip. There are three basic types of guns:
- *Air-cooled guns,* used with a self-shielded electrode, have a small nozzle on the end that does not need to be removed for welding. See **Figure 3-22.**
- *Gas-cooled guns,* used with a gas-shielded electrode, have adapters on the end for the gas nozzle. The nozzles come in different sizes and are usually copper or chrome-plated. See **Figure 3-23.**
- *Water-cooled guns,* used with a large diameter electrode, have a high duty cycle at a high amperage rating, **Figure 3-24.** A water-cooled gun may have a water cooler mounted in a closed system, **Figure 3-25.**

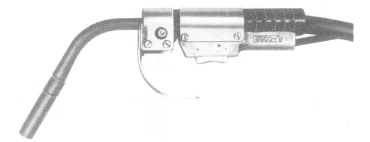

Figure 3-22. This air-cooled gun has an extended nozzle for long stickout welding. The nozzle may be removed for open arc welding. (Lincoln Electric Co.)

Figure 3-24. This radiator cooler is very efficient in lowering water temperatures. (Miller Electric Mfg. Co.)

If city water is used for cooling, a filter and pressure regulator must be used, **Figure 3-26.** City water contains small particles that can clog cooling passages and ruin the welding gun. The filter must be cleaned often, or the filter assembly replaced, to prevent water-flow reduction. City water pressure varies; the regulator prevents excessive pressure. Typically, manufacturers recommend a maximum of 50 psi (345 kPa) in the cooling hoses. Failure to filter the water and regulate pressure will result in clogged guns, burst hoses, and possible damage to the gun.

Mechanized Guns

Mechanized guns are rated for duty cycle just as other types of guns. They contain a 24 V dc drive motor, drive rolls, a small roll of welding electrode, and a feed speed control, **Figure 3-27.** Running on a low-voltage motor, mechanized guns can be used without fear of electrical shock. Depending on the duty cycle of the unit, these guns may be gas- or water-cooled.

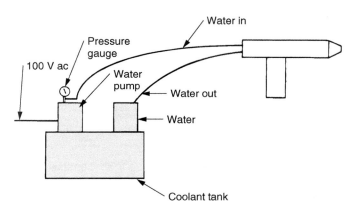

Figure 3-25. A closed system for water-cooling the gun.

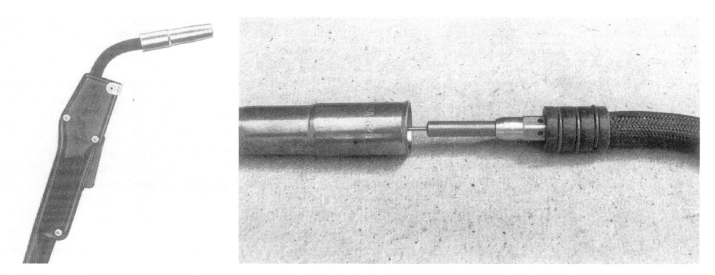

Figure 3-23. The gun and nozzle shown may be used for self-shielded or gas-shielded operation. (Tweco Products, Inc.)

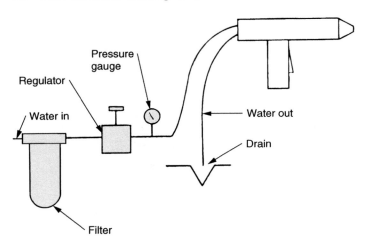

Figure 3-26. City water systems require additional filtering and pressure-regulating equipment to efficiently cool guns.

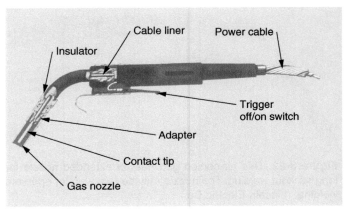

Figure 3-28. Cutaway view of gun components. (L-Tec Welding and Cutting Systems)

Figure 3-27. This motor operates on 24 V dc. The low voltage does not pose a threat of electrical shock. (Miller Electric Mfg. Co.)

Size (Diameter)		No.
3/16	=	No. 3
4/16	=	No. 4
5/16	=	No. 5
6/16	=	No. 6
7/16	=	No. 7
8/16	=	No. 8
9/16	=	No. 9
10/16	=	No. 10
11/16	=	No. 11
12/16	=	No. 12
Lengths		
Short		
Regular		
Long		
Extra Long		
Special		

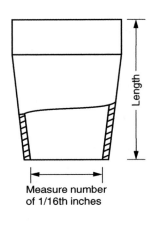

Figure 3-29. Standard nozzle diameters and lengths.

Adapters are available to mate the welding gun to the main feeder. The small roll of electrode is removed from the gun and replaced with a cable assembly from the main feeder, converting the unit to a push-pull system. Refer to Figure 3-11.

Gun Parts

Two principle components of the welding gun are the gas nozzle and contact tip, **Figure 3-28.** All components should fit together firmly.

Gas nozzles direct inert gas flow over the weld area. They are manufactured to fit a specific model of gun and are not interchangeable. FCAW nozzles are either copper or chrome-plated. Exit holes are dimensioned by sixteenths of an inch, **Figure 3-29.**

Contact tips conduct electrical current from the power supply to the consumable welding electrode. When the exit hole enlarges so much that electrical contact is intermittent, the contact tip must be replaced. Continued use of an oversized hole in the contact tip will cause arc outages and current surges to the work. Gun manufacturers specify the proper tip for use with each application. Contact tips are expendable, so several should be kept in stock.

When high welding currents are used, the gas nozzle sometimes produces pieces of copper oxide that can fall into the molten pool. An antispatter compound is used to prevent this condition as well as the formation of spatter from the molten pool in the nozzle opening. Apply the compound inside and outside the nozzle, and on the contact tip and tip adapter, **Figure 3-30.**

Gun Maintenance

A welding gun will operate properly with reasonable maintenance. Treating a gun roughly can damage the outer case and cause internal shortages of electrical current. Do not force threaded assemblies together or use nondesignated parts. When changing adapters or contact tips, clean the threaded areas with a wire brush to ensure

Figure 3-30. Antispatter spray is used to prevent spatter buildup on the end of the nozzle. The compound should be applied often to the nozzle and contact tip. (G.S. Parsons Co.)

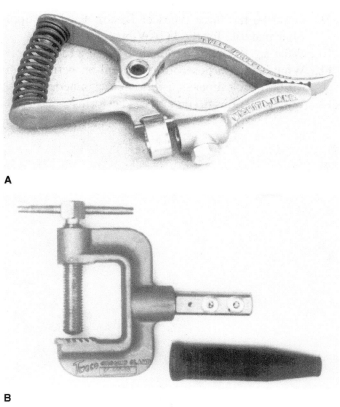

A

B

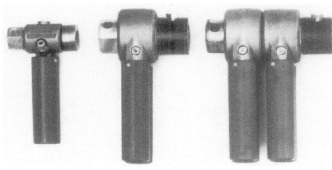

C

Figure 3-31. Ground connections. A—Spring type. B—Screw-clamp type. C—Rotary type. (Tweco Products, Inc.)

electrical contact. Check liner setscrews often, and blow out the liner assembly with compressed air or gas. Always wear a face shield when using compressed air.

Frequently inspect steel liners in the front of the gun assembly for wear and replacement. Guns and feeders for push-pull systems with drive rollers should be checked for proper tension. The feed roller adjustment screw should be tight enough to pull the electrode, but not so tight that the teeth of the wheel leave deep marks in the electrode. Deep indentations will cause small particles of metal to flake off the electrode or split the electrode and wear the gun liner.

Ground Clamps (Work Leads)

Proper clamps and ground leads are extremely important in FCAW. The cable must be adequate enough to carry the welding current without overheating. The ground clamp must fit the workpiece tightly; otherwise, an arc will form where the current flows through the connection, **Figure 3-31.** Arc outages may occur at areas of poor contact, resulting in an unsatisfactory weld. Cables that have cracks in the insulation should be replaced. Connections between the clamp and the cable should be tight. Loose connections occur when the ground clamp is moved frequently, causing the cable to fray at the connection point. To check the ground system, turn off the welding current, and place your hand near the cable and the connectors. If they are warm or hot, servicing or replacement is needed.

Review Questions

Please do not write in this text. Write your answers on a separate sheet of paper.

1. Welding power supplies are controlled by specifications established by the _____.
2. Power supplies are designed to operate for a period of time at a rated load called the _____.
3. In what class is a 150 A welding machine with a 30% duty cycle?
4. The electrical measurement of the arc gap between the end of the electrode and the workpiece is called the _____.
5. What is the function of a constant voltage power supply?
6. When using a constant-current power supply, the _____ control should be set as _____ as possible.
7. Explain how to clean a power supply.
8. Installation of a power supply requiring input power higher than available would require the use of what type of transformer?

9. Describe the three types of feeders used to supply filler material for the FCAW process.

10. What are the major problem areas of a feeder?

11. What is the purpose of the current-on period set by the burnback timer in a welding machine/feeder?

12. What functions do gun cables perform?

13. What is important to know about using city water for cooling?

14. Why can mechanized guns be used without fear of electrical shock?

15. When must contact tips be replaced?

CHAPTER 4 Shielding Gases and Regulation Equipment

Objectives

After studying this chapter, you will be able to:
- State the purposes of shielding gases.
- Describe the properties of shielding gases.
- Explain the various ways shielding gases are distributed to the welding area.
- Explain how gas flow is regulated.

Important Terms

argon
backflow check valve
carbon dioxide
Dewar cylinder
flowmeter
manifold
regulator/flowmeter
shielding gases
Y-valve system

Shielding Gases

Carbon dioxide and argon are the **shielding gases** used in FCAW. Shielding gases have several purposes. They serve to:
- Shield the electrode and the molten metal from the atmosphere.
- Transfer heat from the electrode to the metal.
- Stabilize the arc pattern.
- Aid in controlling bead contour and penetration.
- Assist in metal transfer of the electrode.
- Assist in cleaning the joint.
- Provide wetting action for the filler material.

Carbon Dioxide

Carbon dioxide (chemical symbol CO_2) is a compound gas made up of carbon and oxygen. Although it is not an inert (chemically inactive) gas, it may be used as a primary gas or part of a gas mixture. Since CO_2 dissociates (breaks down) at welding temperatures, the arc is more erratic and harsh, and tends to have more spatter when the gas is used alone. Regardless of whether it is used as a primary or mixed gas, the CO_2 must be *welding grade* gas. Welding grade gas is dried to remove moisture. Welds made with other grades of carbon dioxide may exhibit porosity.

Argon

Argon (chemical symbol Ar) is an inert gas that will not combine with any product of the weld area. It is pure, colorless, and tasteless.

Argon is not used as a primary gas; it is always used in combination with carbon dioxide. Argon is separated from the atmosphere during the production of oxygen, and is readily available at low cost.

Argon produces good bead profiles because the arc is more concentrated than with any other gas. Weld penetration is also good. Spatter and contamination are reduced since the gas is heavier than air and tends to form a blanket around the electrode and molten metal.

Gases and Gas Mixes

The proper gas or gas mixture for FCAW is determined by the electrode manufacturer. Manufacturers add certain elements to the flux to obtain mechanical values, produce specific element percentages, deoxidize weld metal and, in some cases, create a weld with low hydrogen content. To ensure the elements transfer across the arc and maintain these additional characteristics, the prescribed gas must be used. Carbon dioxide is the most popular gas for general welding when spatter is not a problem on the completed weldment. An argon-carbon dioxide mixture with 50% to 75% argon content is also common.

Purge Gas Applications

When it is necessary to purge the root side (backside) of a weld to protect against atmospheric contamination, argon, nitrogen, or helium gases can be used. Argon and helium are used when the application requires very good protection. In many cases, nitrogen can be used as an inexpensive initial purge gas, followed by the more expensive argon or helium. Nitrogen may also be used to protect the backside of fillet or groove welds that do not require full penetration of the root side of the weld. See **Figure 4-1.**

Gas Purity

Inert shielding gases are highly refined. Cylinder argon has a minimum purity of 99.996% and a maximum 15 parts per million (ppm) of moisture (maximum dew point temperature –73°F or –58.3°C). Driox™ (manufactured by Linde Co.) has a minimum purity of 99.998% and a moisture content of less than six ppm. The minimum dew point temperature for welding-grade carbon dioxide is –40°F (–40°C). Some manufacturers produce CO_2 with a moisture content as low as 9 ppm.

The main reason for wanting low moisture is to prevent porosity in the weld. Using contaminated gas, or gas with a high moisture content, leads to additional costs for repairing or reworking the welds. If you suspect that a gas may be contaminated with moisture, send the cylinder to the supplier for inspection.

Gas Supply

Shielding gases are supplied in various styles and sizes of cylinders, **Figure 4-2.** When large quantities of gases are required, ***Dewar cylinders*** (liquefied gas containers also known as Dewar flasks) and high-pressure gas cylinders mounted on trailers are used, **Figure 4-3.** Gas is sold in cubic feet through a distributor.

The distributor charges a demurrage (rental) fee on each cylinder. In areas where large amounts of gas are used, it is more economical to use liquefied gases because fewer cylinders are needed. Many shops will purchase cylinders to avoid the rental fee imposed by the distributor. Purchased cylinders must undergo a hydrotest every five to ten years to ensure their quality and safety.

Storage

Storage of shielding gas cylinders and containers should be strictly controlled to prevent incorrect use. Gas cylinders or containers should always be stored in an outside or well-ventilated area. Cylinders should be kept in an upright position and secured to an upright support. The safety cap must be installed on the cylinder when not in use. Make sure the cylinder is properly labeled to identify the gas contents. Follow all safety precautions to avoid injury. Remember that inert gases do not contain oxygen and, therefore, do not support life. You cannot see, smell, or taste inert gases.

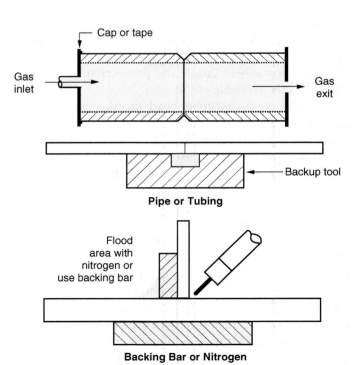

Figure 4-1. Use nitrogen as a purge gas on carbon steels only. Do not use nitrogen for 100% penetration welds.

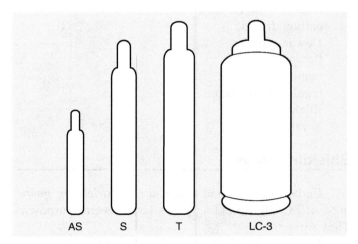

Cylinder Style	Contents (cu. ft.)	Full Pressure of Cylinder at 70°F (psi)	Height (inches)	O.D. (inches)
AS	78	2200	35	7 1/8
S	150	2200	51	7 3/8
T	330	2640	60	9 1/4
LC-3	2900	55	58	20

Figure 4-2. Styles and capacities of cylinders and Dewar flasks supplied to industry.

Gas Distribution

Gases are distributed to the welding area in several ways. Single cylinders are most practical when the gas demand is small. A bank of cylinders is used when a considerable amount of gas is required, **Figure 4-4.** When one bank is emptied, another bank replaces it. Empty cylinders can be replaced without affecting the bank in use.

A *manifold* reduces the number of individual cylinders needed at a welding station, **Figure 4-5.** Shop manifolds can distribute several types of gases to numerous locations, **Figure 4-6.** To supply a manifold, a

Figure 4-3. Two Dewar flasks containing argon are connected to a pressure regulator and a switching valve. (Air Reduction Co.)

Figure 4-5. A six-station manifold attached to a Dewar flask. Dewars supply gas at approximately 50 psi (345 kPa). (Distribution Designs, Inc.)

Figure 4-4. Individual cylinders are connected to a supply manifold by high-pressure tubing called "pigtails." (Air Reduction Co.)

Figure 4-6. Copper tubing is used to carry the major gases to the welding area. This design avoids single tanks in the welding area, the cost of moving cylinders, and safety concerns.

commercial switching arrangement is used to isolate each cylinder, **Figure 4-7.**

The distribution system must be expertly constructed to maintain the purity of the gas in the system. If the main system is made of copper, the joints are usually soldered or silver-brazed. Each tube joint and fitting must be properly cleaned, fluxed, and soldered to allow the complete flow of filler material through the joint. Openings must be capped when not in use. A shut-off valve should be installed when a flowmeter is used, **Figure 4-8.** Use Teflon® tape when connecting any threaded fittings.

Testing for Gas Leaks

Before use, a shielding gas system must be tested for leaks at a pressure higher than the normal operating pressure. One method is to apply a leak-detector fluid or soap to the test area while the pipe or tubing is under pressure, **Figure 4-9.** All leaks, no matter how small, should be repaired before placing the unit in service.

Another leak-testing method is to pressurize the manifold with gas (25 psi to 50 psi). Close the inlet gas valve and note any pressure drop in the manifold. If the pressure drops, a leak is present and must be repaired.

Gas lines should be purged before use to remove flux vapors, air, and moisture. Nitrogen can be used as the initial purging gas. Connect the nitrogen to the manifold, open the valve farthest from the nitrogen supply, and set the gas flow rate between 5 and 10 cubic feet per hour (cfh). Place an analyzer at the open end, and read the oxygen content. When oxygen is low, remove the nitrogen and connect argon to the system. Set the flow rate at 5 cfh, and purge until the analyzer shows the system is clear of oxygen.

Purge replacement hoses for several minutes before installation. Do not use any hose for the gas supply system that has been used for fluids.

Gas Regulation

A *regulator* reduces the cylinder pressure to the desired manifold pressure level (refer to Figure 4-7). Pressure is usually set between 20 psi and 50 psi (138 kPa and 345 kPa). A *flowmeter* at the welding station regulates the flow of gas to the welding gun, **Figure 4-10.**

A *regulator/flowmeter* is used to reduce cylinder pressure *and* control the amount of gas to the gun. The tank outlet and regulator fitting should be free of nicks

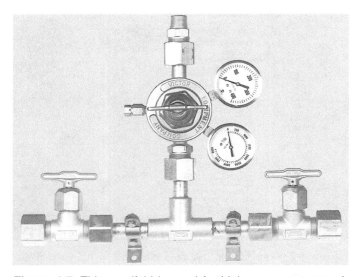

Figure 4-7. This manifold is used for high-pressure gases. A regulator is attached to reduce cylinder pressure to the desired manifold pressure. (Victor Equipment Co.)

Figure 4-8. Shutoff valves are used on the end of the low-pressure manifold so the flowmeter may be removed or changed.

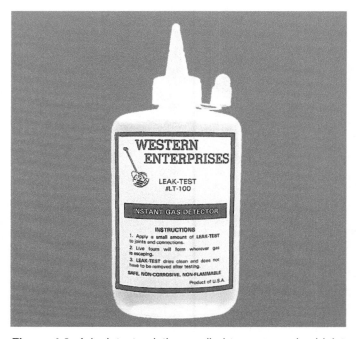

Figure 4-9. A leak-test solution applied to a pressurized joint will bubble if a leak is present. (Western Enterprises Co.)

Figure 4-10. This station flowmeter operates on manifolds at approximately 50 psi (345 kPa). (Linde Co.)

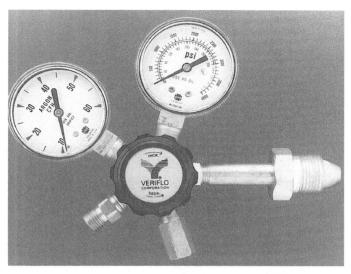

Figure 4-12. A dial gauge is used with a regulator to read gas flow in cubic feet per hour. (Veriflow Co.)

and dirt. Argon regulators use a ball taper connection for sealing the regulator to the cylinder, **Figures 4-11** and **4-12.** The fittings on the CO_2 cylinder and regulator are flat and require a washer to prevent leaks, **Figure 4-13.** If leaks occur during a soap test, the washer must be replaced.

Figure 4-13. The CO_2 regulator must be attached to the cylinder with a flat washer to prevent leakage. (Union Carbide)

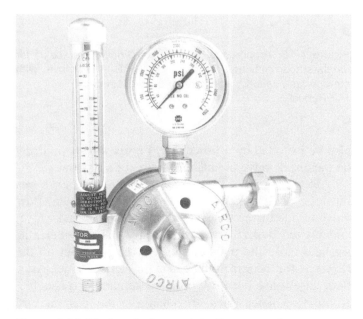

Figure 4-11. Single cylinder regulator/flowmeters have a gauge that shows cylinder pressure. Gas flow to the gun is controlled by turning the adjustment knob and reading the ball position in the vertical tube. (Air Reduction Co.)

When installing a regulator/flowmeter or a flowmeter with a ball tube, keep the tube in the vertical position. The amount of flow is indicated at the top of the ball, unless otherwise indicated. The tube must be calibrated for use with a particular gas and cannot be used with any other gas or mixture.

Regardless of the type of gas supply (cylinder, Dewar flask, manifold), when the gas flow valve is opened, a surge of built-up gas exits the nozzle for several seconds. A surge check valve can be used to eliminate this condition, **Figure 4-14.**

Gas Mixing

When large volumes of gases are used, mixing can be done at the manifold, **Figure 4-15.** Single- or multiple-station installations are also available, **Figure 4-16.**

Figure 4-14. Surge check valves eliminate gas surging from the nozzle during the start of operation. They quickly pay for themselves in gas savings. (Weld World Co.)

A

B

Figure 4-16. Proportional mixers. A—Small mixer used for single-station mixing. (Tescom Corp.) B—Mini-mixer with storage tank used for individual or manifold stations. (Air Reduction Co.)

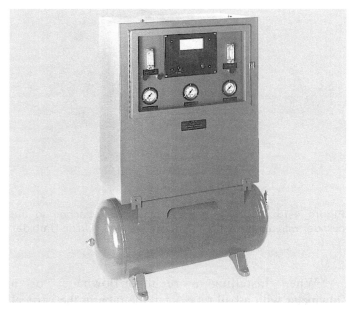

Figure 4-15. The large tank on the bottom of the mixer serves as a mixing tank and storage chamber. (Thermco Instrument Co.)

The **Y-valve system** is often used in single welding stations to change gases that are different from standard mixes. Y-valves are installed on the outlet side of the flowmeter, with the gas metered by two separate flowmeters, **Figure 4-17.**

To prevent backflow and improper gas mixing, a **backflow check valve** is installed between the flowmeter and Y-valve, **Figure 4-18.** A backflow check valve can also be installed on a flowmeter, **Figure 4-19.** An arrow on the check valve indicates the direction of flow.

To properly set up the gas flow mixture, each gas must be set independently. Open the main gas flow valve to one of the gases. Open one side of the Y-valve, and set the desired flow rate. Leave the flowmeter as set, and close the main supply for that gas. Set up the flow rate for the remaining gas in the same manner. After both gases have been set, open the remaining gas valve. If the flowmeter rates are changed, the gas mixture percentages will change, and the system will require resetting.

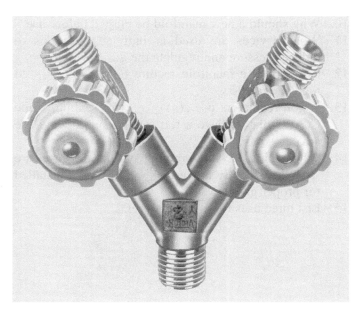

Figure 4-17. Y-valves have shutoff valves for the single flow of one gas or a mixture of two gases. (Victor Equipment Co.)

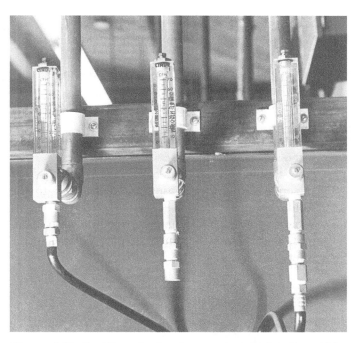

Figure 4-19. Backflow check valves are installed on the middle and right flowmeters.

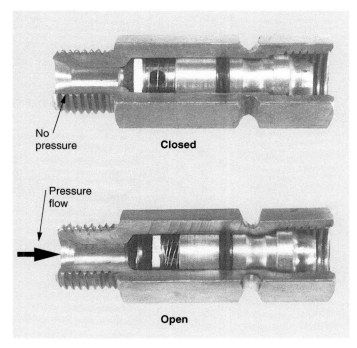

Figure 4-18. Backflow check valves prevent the mixing of gases in the supply line. (Air Reduction Co.)

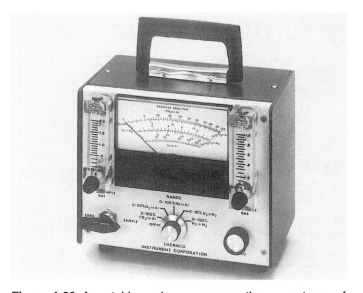

Figure 4-20. A portable analyzer measures the percentages of individual gases. (Thermco Instrument Co.)

Mixture and Purity Testing

Analyzers can be used for testing gas mixtures at the welding station, **Figure 4-20.** Gas analyzers can also be used to check for leaks in the gas supply system, to test the purity of the gas being used, and to check for adequate purging of pipes and vessels before welding.

Review Questions

Please do not write in this text. Write your answers on a separate sheet of paper.

1. Which is *not* a purpose served by shielding gases?
   A. Protection from the atmosphere.
   B. Transfer heat from the electrode to the metal.
   C. Promote oxidation.
   D. Stabilize the arc pattern.

2. What shielding gases are used for FCAW?
3. Why do welds made with carbon dioxide have more spatter?
4. What gas is often used as an initial purging gas because of its low cost?
5. What is the result of welding with a shielding gas that has a high moisture content?
6. What is the name for liquefied gas containers, and when are they used?
7. How should gas cylinders be stored?
8. Describe three ways gases are distributed to the welding area.
9. What two tests should be performed on shop manifolds to maintain the purity of the system and the safety of the user?
10. Why should a new manifold be purged before use?
11. What devices are used at individual stations to reduce pressure and regulate the gas flow rate?
12. Gases from a manifold require only a(n) _____ to regulate gas flow.
13. Argon regulators use a(n) _____ for sealing, while CO_2 regulators have a flat sealing surface and use a(n) _____.
14. When using a Y-valve mixing system, the gases are set independently and a(n) _____ must be installed for proper gas flow.
15. List three uses for a gas analyzer.

5 *Filler Material*

Objectives

After studying this chapter, you will be able to:
- Differentiate types of FCAW electrodes.
- Describe the forms in which filler material is provided.
- Properly handle filler material to prevent its contamination.
- Indicate the factors affecting selection of the proper electrode.

Manufacturing FCAW Electrodes

In FCAW, a tubular electrode and an internal flux are used to deposit the weld metal. The flux mixture consists of deoxidizers, slag formers, and arc stabilizers. The electrode base material is formed into a "U" shape, flux is added, and a tube is formed. The tube is drawn (reduced) to a smaller size, compressing the flux into a tight matrix, **Figure 5-1.**

The chemical composition of the electrode and flux can be tightly controlled to yield an exact composition in the weld. Since the flux is sealed in the electrode tube, storage life is extended and handling problems are diminished.

Defects commonly found on the surface of a solid electrode are not evident on a tubular electrode. The tube is drawn from thin sheet metal; thus, the reduction area is much less than when drawing from a solid material. After the tube is formed, it is heat-treated (if required), cleaned, and packaged for final use.

Filler Material Specifications

The AWS issues specifications for FCAW tubular electrodes used with carbon steels, low-alloy steels, corrosion-resisting chromium, and chromium-nickel steels. Other electrodes made for surfacing, cladding,

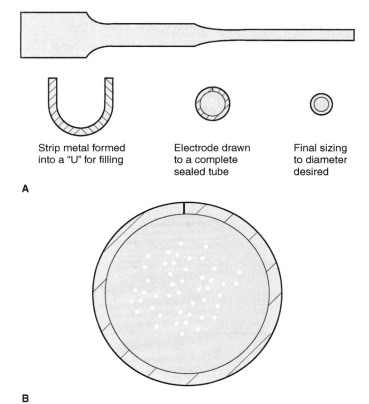

Strip metal formed into a "U" for filling

Electrode drawn to a complete sealed tube

Final sizing to diameter desired

A

B

Figure 5-1. Manufacture of the tubular electrode. A—Metal is formed into a "U," filled with core material, drawn into a tube, and sized to a precise diameter. B—Cross section of core material, consisting of powdered metals, vapor-forming material, deoxidizers, scavengers, and slag formers.

and repairing cast iron are made to a commercial or AWS specification. AWS specifications for steel and stainless steel are:
- *AWS A5.20 Carbon Steel Electrodes for FCAW.*
- *AWS A5.22 Stainless Steel Electrodes for FCAW.*
- *AWS A5.29 Low Alloy Steel Electrodes for FCAW.*

A number following the specification number indicates the year of publication. Always refer to the most recent specifications available. Carbon steel and low-alloy steel electrodes are identified by a coded system, **Figure 5-2**. Chromium and chromium-nickel steels use a base-metal system for identification, **Figure 5-3**. Additional information for flux cored electrodes is provided in the Reference Section.

Many electrodes are manufactured for other purposes, such as joining, cladding, hardfacing, or building up welds. They are made to company or special specifications that govern the chemical content, mechanical properties, and usability of the filler material.

To ensure all electrodes meet equal standards of quality, AWS specifications establish requirements for

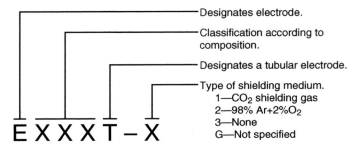

Figure 5-3. AWS classification of FCAW chromium and chromium-nickel steel electrodes.

scope, classification, acceptance, manufacture, and testing. Additional requirements may include chemical tests, impact test results, and special packaging.

Electrode manufacturers guarantee only that their products meet specifications. They will replace defective electrodes but will not guarantee acceptable results, since they cannot control the welding process.

Filler Material Form

Filler material is wound on 4″, 8″, or 12″ diameter spools or coiled inside large drums. Check your machine to determine the correct spool or coil diameter. Spools are made of plastic, wood, or formed metal and are disposable, **Figure 5-4**. Coils vary in weight and size, **Figure 5-5**. Electrodes range in diameter from 0.030″ to 5/32″, with tolerances of 0.002″ to 0.003″. Material is wound on spools and into coils in a random manner called layer wound, **Figure 5-6**.

To prevent feeding problems, electrodes must meet the following requirements:

- Each coil, with or without support, spool, or drum, must contain one continuous length of electrode made from a single lot of material.
- Butt welds, when present, must not interfere with the uniform, uninterrupted feeding of the electrode on automatic and semiautomatic equipment.
- The electrode must be able to unwind freely without being restricted by kinks, waves, or sharp bends.

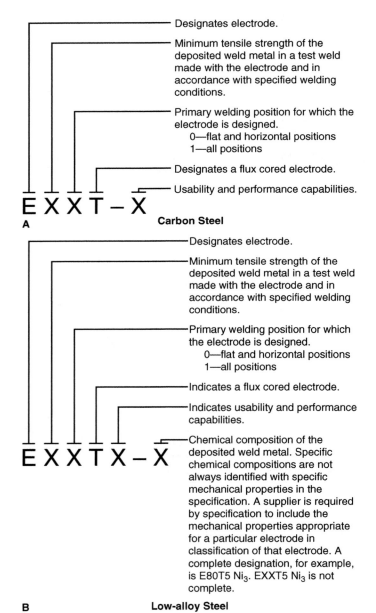

Figure 5-2. AWS classification of FCAW electrodes. A—Carbon steel. B—Low-alloy steel.

Figure 5-4. Various types of spools for standard electrode feeders.

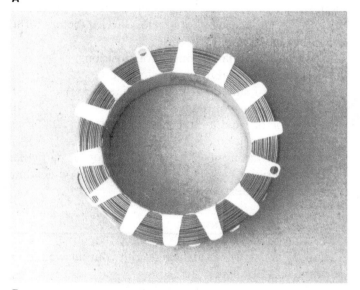

Figure 5-5. Coils. A—The electrode is wound on a cardboard backing. B—The electrode is wound on a metal frame. The type of material, heat code, size, and other descriptive information are printed on the frame.

Figure 5-6. A layer wound spool of material mounted on a spool.

Identification and Packaging

Information labels are located on the outside of the spool flange, **Figure 5-7,** or inside the coil liner, **Figure 5-8.** Drums storing coiled electrodes are also labeled, **Figure 5-9.**

Electrodes are packaged in waterproof boxes, tin cans, and plastic bags for protection from contamination, **Figure 5-10.** Filler material is safe from contamination as long as it remains in a sealed package.

Using Filler Material

Preventing contamination of filler material after the package is opened is the responsibility of the user. Electrodes are easily contaminated by contact with oil, grease, moisture, soot, soiled rags, or a dirty work area,

Figure 5-7. The electrode is wound on a plastic spool. The upper label provides data for the type of electrode. The lower label provides safety information for use of the material.

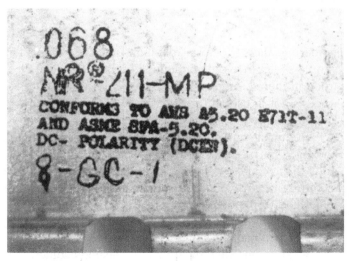

Figure 5-8. Electrode data is printed on the steel backing material around which the material is wound.

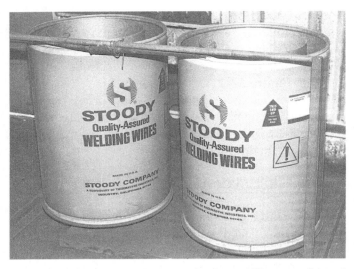

Figure 5-9. Drums contain large amounts of filler material and are used in surfacing or cladding operations.

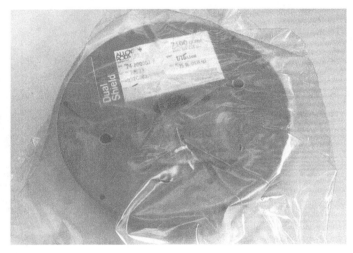

Figure 5-10. A spooled electrode sealed in a plastic bag will store indefinitely.

rendering them useless for high-quality welds. The simplest way to avoid contamination is to keep the electrode in the package as long as possible. When an electrode is not in use, put it back in the original package to keep it clean.

Requiring manufacturers to clean, inspect, and package the filler material to specifications does little good if it is improperly handled and contaminated before or during use. Several practices will help keep the filler material uncontaminated:

- Keep materials in their original packaging.
- Open packages only when needed.
- Store unsealed materials in a heated cabinet.

- Handle materials as little as possible. Wear clean gloves when you do handle them.
- Remove spools and coils from machines when they are idle for an extended period of time.
- When changing spools or coils, inspect the drive rollers, guides, tips, and other components for wear and possible replacement.
- When changing electrodes, inspect drive rollers, guide tubes, liners, and contact tips for metal particles. Any metal particles must be removed; otherwise, the new filler material may be contaminated.

Selecting Filler Material

One of the most important considerations in FCAW is selecting the correct electrode. The electrode, in combination with the proper shielding gas, produces the deposit chemistry that determines the physical and mechanical properties of the weld. Several factors influence the choice of electrode:

- Base material chemical composition.
- Base material mechanical properties.
- Type of shielding gas used (if required).
- Service use of the weldment.
- Type of weld joint design.
- Weld position.

The selection of filler material for a particular welding operation is discussed in the chapters on welding specific base materials.

Review Questions

Please do not write in this text. Write your answers on a separate sheet of paper.

1. How is filler material developed?
2. Why does a tubular electrode have fewer defects than a solid electrode?
3. Many of the filler materials used in FCAW are made to whose specifications?
4. In the classification E80T5 Ni_3, what does the letter E designate? What does Ni_3 designate?
5. *True or false?* Electrode manufacturers guarantee the product and the results.
6. In what two forms is filler material provided?
7. What are the sizes and tolerances for coiled electrodes?
8. How are electrodes packaged?
9. Why is it important to protect electrodes from contamination?
10. List six factors influencing the selection of the proper electrode.

6 Weld Joints and Weld Types

Objectives

After studying this chapter, you will be able to:
- Identify the basic types of weld joints.
- Recognize the types of welds made for each type of joint.
- Identify some modified weld joint configurations.
- Interpret introductory welding terminology and symbols.
- Describe how weld designs and joints are evaluated.

Important Terms

buttering
interpass temperature
joint
mismatch
postheating
preheating
stringer beads
surfacing
tooling
welding symbol

Joint Types

A *joint* is the junction or edges of members that are joined. There are five basic types of joints, **Figure 6-1:**
- Butt joint.
- T-joint.
- Lap joint.
- Corner joint.
- Edge joint.

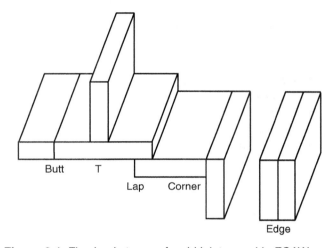

Figure 6-1. Five basic types of weld joints used in FCAW.

Weld Joint Preparation

Weld joints may be initially prepared in a number of ways, including:
- Shearing.
- Casting.
- Forging.
- Machining.
- Stamping.
- Filing.
- Routing.
- Oxyacetylene cutting (thermal cutting process).
- Plasma-arc cutting (thermal cutting process).
- Grinding.

The final preparation of the joint before welding will be discussed in subsequent chapters covering the welding of a particular metal.

Weld Types

Various types of welds can be made for each of the five basic joints.

Welds made for *butt joints,* **Figure 6-2,** include:
- Square-groove butt weld.
- Bevel-groove butt weld.
- V-groove butt weld.
- J-groove butt weld.
- U-groove butt weld.
- Flare-V-groove butt weld.
- Flare-bevel-groove butt weld.

Welds made for *T-joints,* **Figure 6-3,** include:
- Bevel-groove weld.
- J-groove weld.
- Fillet weld.
- Slot weld.
- Plug weld.
- Flare-bevel-groove weld.

Welds made for *lap joints,* **Figure 6-4,** include:
- Fillet weld.
- Bevel-groove weld.
- J-groove weld.
- Plug weld.
- Slot weld.
- Spot weld.

Welds made for *corner joints,* **Figure 6-5,** include:
- Butt weld or square-groove weld.
- Fillet weld.
- J-groove weld.
- Spot weld.
- V-groove weld.
- Edge weld.

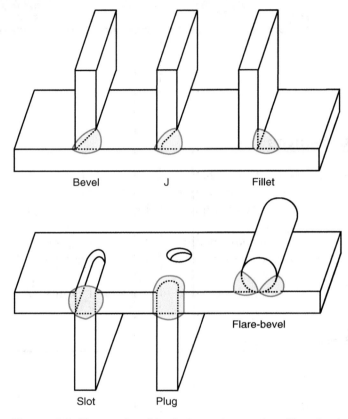

Figure 6-3. Types of welds that can be made with a basic

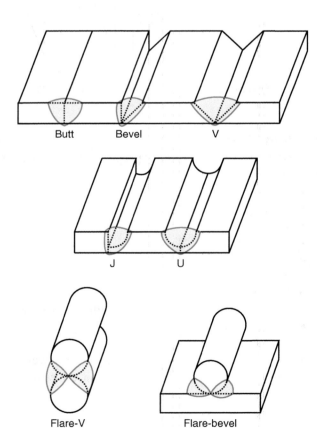

Figure 6-2. Types of welds that can be made with a basic butt joint.

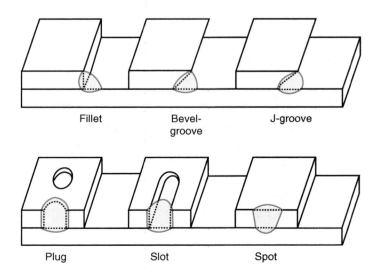

Figure 6-4. Types of welds that can be made with a basic lap joint.

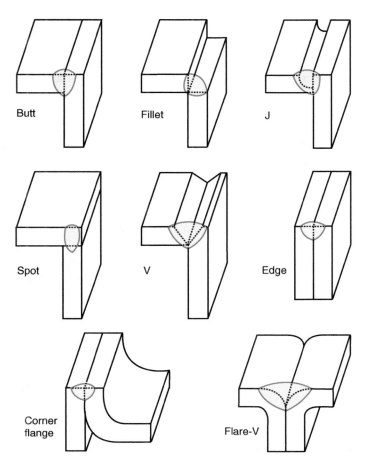

Figure 6-5. Types of welds that can be made with a basic corner joint.

- Corner-flange weld.
- Flare-V-groove weld.

Welds made for *edge joints,* **Figure 6-6,** include:
- Butt weld or square-groove weld.
- Bevel-groove weld.
- J-groove weld.
- V-groove weld.
- U-groove weld.
- Edge-flange weld.
- Corner-flange weld.

Double Welds

Many groove welds are made from both sides of the joint. The advantages of double-groove welds are:
- Less distortion of the final weldment.
- Less weld metal is needed, lowering cost.
- Preparation cost may be lower.
- Complete penetration through the joint.

Many fillet welds are also made from both sides of the joint. The advantages of double-fillet welds are:
- Less distortion of the final weldment.
- Smaller fillet welds without loss of strength.
- Added protection for joints that contain fluids.

Applications for double welds are illustrated in **Figure 6-7.**

Weldment Configurations

Basic weld-joint configurations are often modified to assist in the assembly of the component parts, improve joint access, or change the metallurgical and physical properties of the weld. Some common weldment designs are described here.

Joggle-type joints are used in tanks (cylinder-to-head assemblies) where backing tooling or backing bars are not effective or cannot be used, **Figure 6-8.** The automotive industry uses this joint type in the manufacture of unibody automobiles where one side of the panel must be flush. One side of the panel is joggled with special joggle tools. The unit is assembled, clamped, and the welding is completed as required.

The *tubular butt joint with built-in backing bar* design is used in tubular assemblies where tooling cannot be inserted into the pipe diameter or where overall dimensions are precise. In this case, the pipes are assembled

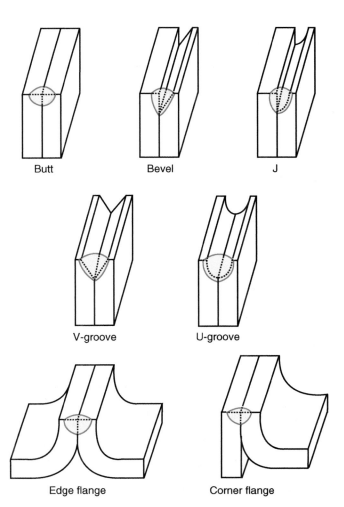

Figure 6-6. Types of welds that can be made with a basic edge joint.

until the lands (root faces) meet, controlling the overall length dimension. **See Figure 6-9.**

Another weldment design is the *butt joint with pre-fabricated backing bar.* The bar may be 1″ flat stock or larger, depending on the thickness of the base material. As the bar is assembled into the cylinder, tack welds are placed on one edge. After assembly of the mating part, the other side of the bar is tack welded. Since the weld usually penetrates into the backing bar, the bar is not removed after welding, **Figure 6-10.**

The major problem with this design is the fitup of the bar to the back of the base material. Any area that does not contact the back of the material will not control the heat flow, and penetration into the backing bar may not occur. Specially designed joints are used for controlling penetration into the joint where excessive penetration could cause a problem with liquid flow. See **Figure 6-11.**

Joining dissimilar metals often requires *buttering*—depositing a material on one of the joint pieces to make the joint materials metallurgically compatible with a common filler material. See **Figure 6-12.**

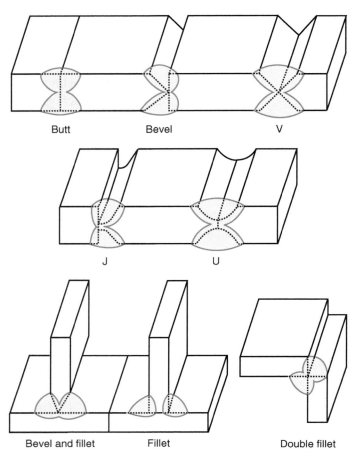

Figure 6-7. Applications for double welds.

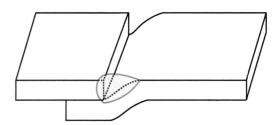

Figure 6-8. Joggle-type joint.

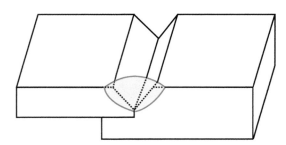

Figure 6-9. Tubular butt joint with a built-in backing bar.

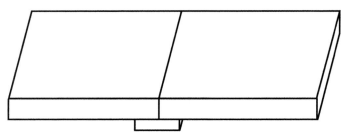

Figure 6-10. Plate butt weld with fabricated backing bar.

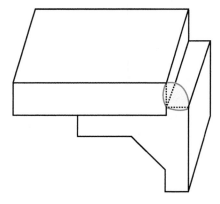

Figure 6-11. Controlled weld penetration joint.

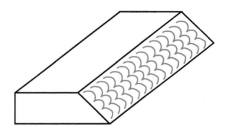

Figure 6-12. Buttered weld joint face.

When a material is required to protect the base metal from chemicals, heat, abrasion, or other forms of wear, a layer of weld metal is placed on the surface. Overlaying a weld in this manner is called *surfacing* or *cladding,* **Figure 6-13.**

Pipe joints often require a special backing ring for assembling and welding the joint. The type of material flowing through the pipe influences the ring's design. Because the weld metal penetrates the backing ring, it is not removable once welding is completed. See **Figure 6-14.**

Welding Terms and Symbols

Communication between the designer and the welder is essential in the welding operation. Weld and weld joint terminology must be mutually understood, **Figure 6-15.** The *welding symbol* designated by the American Welding Society is used as a standard throughout the industry, **Figure 6-16.** The symbol is used on drawings to indicate type of weld, weld joint, and weld placement. The welding symbol may also include information regarding finish contours and testing.

Part of the welding symbol is the *weld symbol.* It is important to study and understand each part. **Figure 6-17** shows basic weld symbols, which direct the welder to select the proper weld joint. The flat line on which a symbol is placed is called the *reference line.* An arrow on one end points to the weld joint and indicates where the weld should be made. The line to the arrow is always at an angle to the reference line. Whenever the weld symbol is placed *below* the reference line, the weld is to be made on

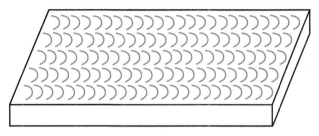

Figure 6-13. Overlay welds protect the base material from wear or contamination.

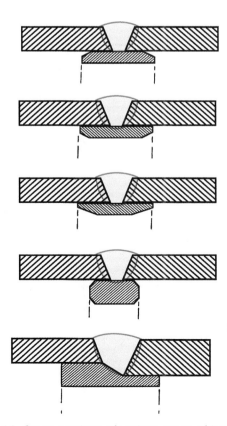

Figure 6-14. Cross sections of various types of backing rings for pipe joints.

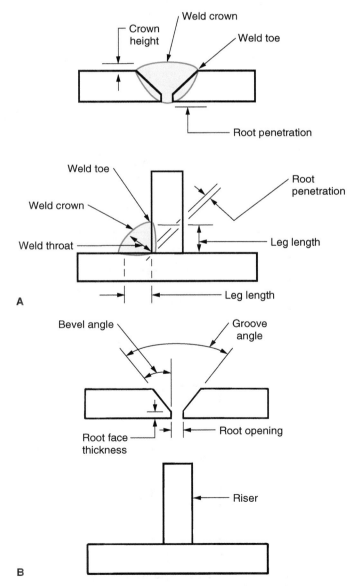

Figure 6-15. Welding terminology. A—For the weld. B—For the weld joint.

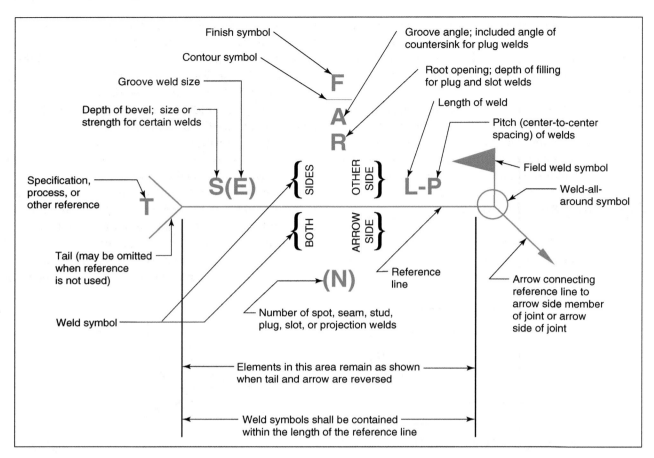

Figure 6-16. The AWS welding symbol conveys specific and complete information to the welder. (From ANSI/AWS A2.4-98. Printed with Permission of the American Welding Society)

Groove							
Square	Scarf	V	Bevel	U	J	Flare-V	Flare-bevel

Fillet	Plug or slot	Stud	Spot or projection	Seam	Back or backing	Surfacing	Edge

Figure 6-17. Basic weld symbols. (From ANSI/AWS A2.4-98. Printed with Permission of the American Welding Society)

the *near (arrow) side*, **Figure 6-18.** Whenever the weld symbol is placed *above* the reference line, the weld is to be made on the *other side* of the joint, **Figure 6-19.** By placing dimensions on the symbols, the exact size of the weld can be indicated. The complete welding symbol gives the welder instructions for preparing the base material, welding process, weld surface contour, and finish. See **Figure 6-20.**

Weld Positions

Many weldments are made on site, so the welder must be able to weld in all positions—flat, horizontal,

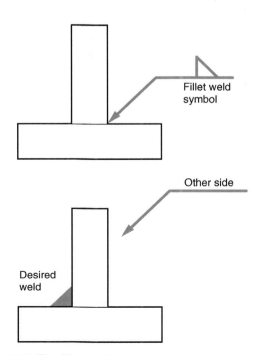

Figure 6-18. The fillet weld symbol shown below the reference line indicates the weld is located on the side at which the arrow points.

Figure 6-19. The fillet weld symbol shown above the reference line indicates the weld is located on the side of the joint opposite where the arrow points.

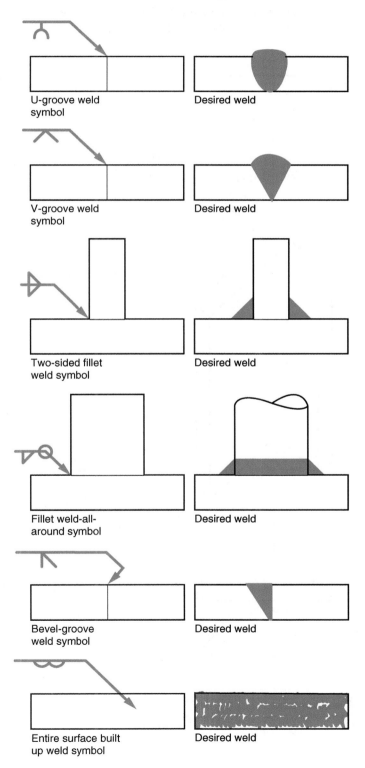

Figure 6-20. Typical weld symbols and applications.

vertical, and overhead, **Figure 6-21.** The effect of gravity on the molten weld pool differs with each position. Heat distribution also varies. These two factors make the skills for each position distinct. Practice is necessary to produce good welds in all positions. There are several considerations when welding in the various positions:

- The *flat* position has the greatest deposition rate. Flat welds usually have less porosity because the gas can rise to the top of the weld pool and escape before the metal solidifies.

- Undercut at the top of the weld pool can be a major problem in the *horizontal* position.
- Heat rises during *vertical* welding, requiring close observation of the molten pool to prevent sagging and overheating of the weld.
- *Overhead* welding is the most tiring of all positions. It also has the slowest metal deposition. Making wash beads is extremely difficult due to metal sagging. Overhead welds are prone to porosity.

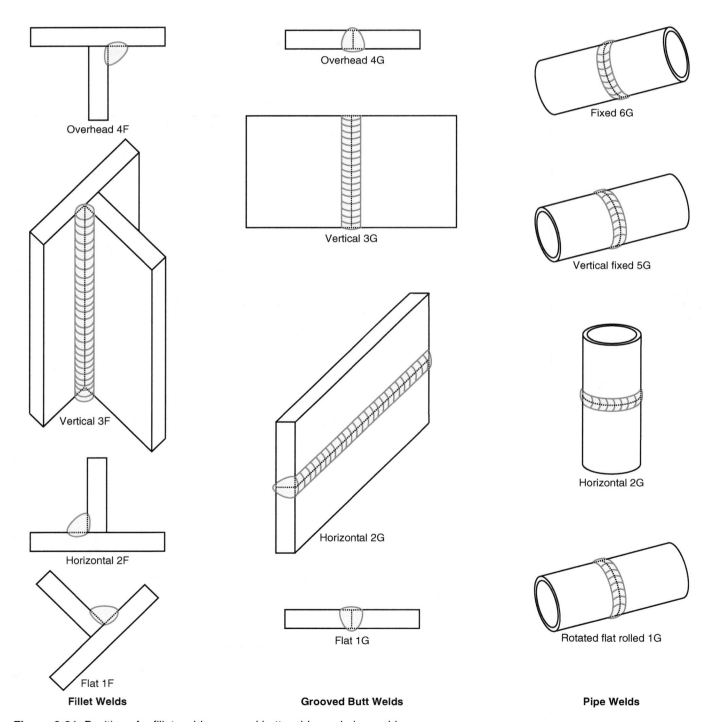

Fillet Welds **Grooved Butt Welds** **Pipe Welds**

Figure 6-21. Positions for fillet welds, grooved butt welds, and pipe welds.

Caution: Wear protective clothing and leathers to shield yourself from falling sparks and metal. Do not stand directly under the spark stream. Wear protective eye lenses. Never clean the slag on a weld without eye protection. Stay aware of your environment when your hood is down and you are welding. If flammable material is present, take measures to prevent fires. Keep a fire extinguisher at your work area.

Design Considerations

The types of weld designs and joints should be evaluated to ensure the weldment does the intended job. Areas of evaluation include:
- Material type and condition (annealed, hardened, tempered).
- Service conditions (pressure, chemical, vibration, shock, wear).
- Physical and mechanical properties of the completed weld and heat-affected zones.
- Preparation and welding costs.
- Assembly configuration and weld access.
- Welding equipment and tooling.

Butt Joint and Welds

Butt joints are used when strength is required. They are reliable and can withstand stress better than any other type of weld. To achieve full stress value, the weld must have 100% penetration. This can be achieved by completely welding through from one side. The alternative is to weld both sides with the weld joining in the center.

Thinner gauge materials are more difficult to fit up for welding. They require costly tooling to maintain proper joint configuration. Tack welding may be used as a method of holding the components during assembly. However, tack welds present several problems:
- Conflicting with the final weld penetration into the weld joint.
- Adding to the final crown dimension (height).
- Cracking during the welding operation due to heating and expansion of the joint.
- Contaminating the weld joint when cleaning the tack weld slag, causing weld defects.

The expansion of the base material during welding can result in *mismatch*, **Figure 6-22.** When mismatch occurs, the weld will not completely penetrate the joint. Many specifications limit highly stressed joints to a maximum mismatch of 10% of the joint thickness.

Whenever possible, butt joints should mate at the bottom of the joint, **Figure 6-23.** Joints of unequal thicknesses should be machined in the weld area to provide even surfaces for adequate fusion, **Figure 6-24.** Where this cannot be done, the heavier piece may be machined on the upper part of the joint.

Weld Shrinkage

Butt welds always shrink across (transverse) the joint during welding. A shrinkage allowance must be made if the post-welding dimensions have a small tolerance, or if the outer edges of the material are to be trimmed to a final dimension. Butt welds on pipe, tubing, and cylinders also shrink, **Figure 6-25.** When dimensions

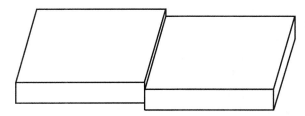

Figure 6-22. When under stress, welds made with mismatched joints often fail below the rated load.

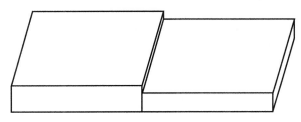

Figure 6-23. Mating the joint at the bottom equalizes the load under stress. The weld is placed on the top and penetrates completely through the joint.

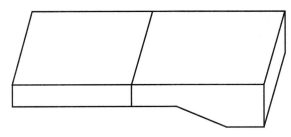

Figure 6-24. Joints of unequal thickness absorb different amounts of heat and expand at different ratios. Equalize heat flow by tapering the heavier material to the thickness of the thinner material.

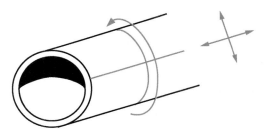

Figure 6-25. During welding, butt welds shrink both transversely and longitudinally.

must be held, a shrinkage test can be done, **Figure 6-26.** Additional material can be added to the overall length of the part for final trimming. Heavier materials will shrink more than thinner materials. Double-groove welds will shrink less than single-groove welds, because less welding is involved and less filler material is used to fill the joint.

Lap Joints and Welds

Lap joints require very little preparation and may be single-fillet, double-fillet, plug-, slot-, or spot-welded. They are used in static-load applications or the repair of unibody automobiles. When corrosive liquids are used, both edges of the joint must be welded, **Figure 6-27.** One of the major problems with lap-joint design occurs when the component parts are not in close contact, as they should be for a fillet weld, **Figure 6-28.** A bridging fillet weld that leads to incomplete fusion at the root of the weld and an oversize fillet weld dimension must be made. When using this type of design in sheet or plate material, adequate clamps or tooling must be used to maintain contact of the material at the weld joint.

In the assembly of cylindrical parts, an interference fit of the mating parts eliminates the problem, **Figure 6-29.** The inside diameter of the outer part is made several thousandths of an inch smaller than the outside diameter of the inner part. Before assembly, the outer part is heated until it expands enough to slide over the inner part. As it cools, it locks onto the smaller part. Small tack welds may be used to hold the assembly together during welding.

T-joints and Welds

T-joints are used for joining parts at angles to each other. T-joints may be single-fillet, double-fillet, or

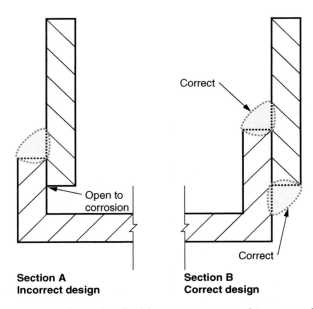

**Section A
Incorrect design**

**Section B
Correct design**

Figure 6-27. Corrosive liquids must not enter the penetration side of the weld joint. In Section B, the back of the weld is closed to corrosion. In Section A, the back of the weld is open to corrosion.

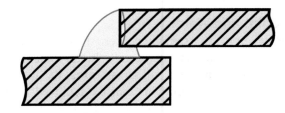

Figure 6-28. Lap joint problem areas are often the result of improper fitup.

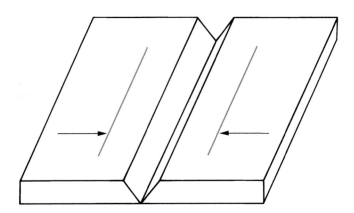

1. Tackweld the test joint together.
2. Scribe parallel lines, as shown, with approximately 2″ centers. Record this dimension.
3. Weld the joint with the test weld procedure.
4. Measure the scribed lines and compare it with the original dimension.

Figure 6-26. Determining weld joint shrinkage.

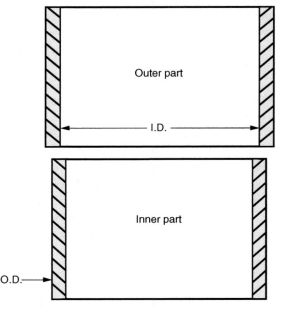

Figure 6-29. In an interference fit of mating cylindrical parts, the ID of the outer part is made several thousandths of an inch smaller than the OD of the inner part.

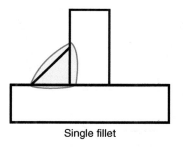
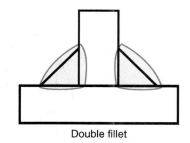
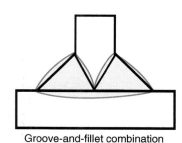

Figure 6-30. Types of T-joints and welds.

groove-and-fillet welds, **Figure 6-30.** Fillet weld sizes must conform to allowable design loads. AWS specifies where fillet welds may be used, as well as the minimum and maximum sizes permissible in the construction of buildings and bridges. When welding on materials less than 1/4″ thick, and design loads are unknown, follow this rule of thumb: the fillet weld lengths must equal

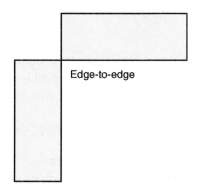

Figure 6-31. Fillet weld legs should be equal in length from the root of the joint. Unequal leg length, unless specified, will not carry the designed load and may fail when stressed.

the thickness of the thinner material being joined. See **Figure 6-31.**

The main problem in making fillet welds is lack of penetration at the joint intersection. To prevent this condition, make ***stringer beads*** (welds made without side-to-side motion) at the intersection. Weave beads across the intersection are prone to lack of penetration.

Corner Joints and Welds

Corner joints are similar to T-joints. They consist of sheets or plates mating at an angle to one another, **Figure 6-32.** Corner joints are usually used in conjunction with groove welds and fillet welds. When using thin materials, the assembly of component parts may be difficult without proper tooling. Tack welding and welding without extensive tooling often causes distortion and buckling of thinner materials. For the most part, these types of joints should be limited to heavier materials in structural assemblies.

Edge Joints and Welds

Edge welds are used when the edges of two sheets or plates are adjacent and have approximately parallel planes at the point of welding. See **Figure 6-33.** Edge welds are applicable on thinner gauges of metal and are not used in structural assemblies. Only full penetration welds should be used with stress or pressure applications.

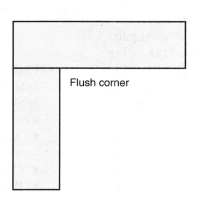
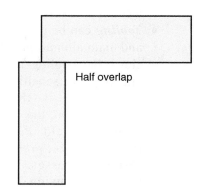

Figure 6-32. Common corner-joint designs used in the fabrication of component parts.

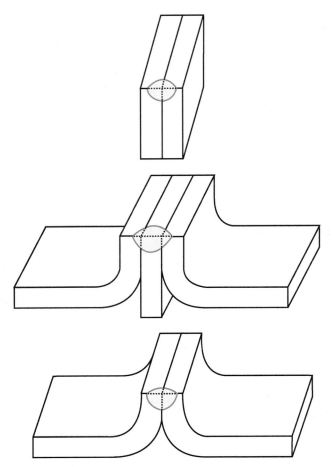

Figure 6-33. Common edge-joint designs.

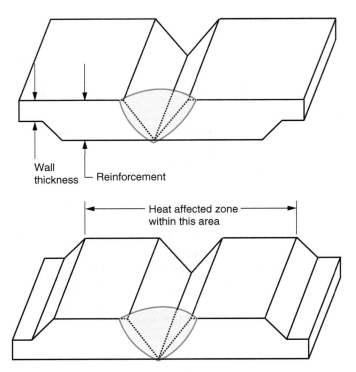

Figure 6-34. Joint thickness and filler material tensile strength combined are equivalent to the strength of the base material.

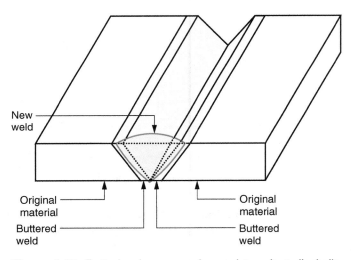

Figure 6-35. Buttering is commonly used to adapt dissimilar metals for welding.

Special Designs and Procedures

Special designs are used in the fabrication of a weldment when:

- Welds cannot be thermally treated due to their configuration or size. Additional material is added to the joint thickness, and the weld is made restricting the heat flow into the thinner material. See **Figure 6-34.** This type of joint will achieve the full mechanical values of the base material.
- Joining dissimilar metals often requires buttering one material to allow the joint to be welded with a common filler material. See **Figure 6-35.**
- *Tooling* can be used to preheat component parts and maintain interpass temperatures. *Preheating* slows the cooling rate of the weld to prevent cracking of sensitive components. *Interpass temperature* is the maximum temperature allowed when the next pass is started on a multipass weld. *Postheating* is used to control grain size, prevent cracking, and remove residual stresses. These temperatures may be checked using special crayons, paints, or pellets, **Figure 6-36.**
- Chill (backing) bars and tooling may be used to localize and remove heat during the welding application. See **Figure 6-37.** Other benefits of tooling include:
  - Aligning components before and during the welding operation.
  - Checking component parts for proper fitup.
  - Preventing excessive drop-through penetration on the other side of the weld.

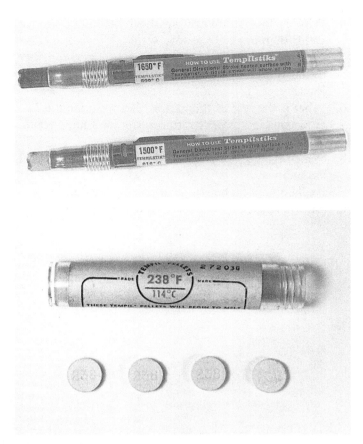

Figure 6-36. Temperature crayons and pellets can be used to determine preheat and postheat temperatures.

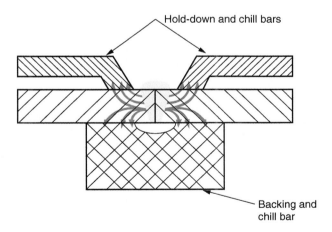

Hold-down and chill bars

Backing and chill bar

Figure 6-37. Tooling and chill bars are used to remove heat from the weld and resist heat flow into the base material.

Review Questions

Please do not write in this text. Write your answers on a separate sheet of paper.

1. Name the five basic types of weld joints.

2. Identify the type of joint and the type of weld shown for each figure.

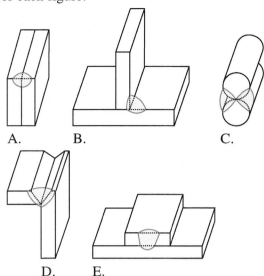

A.   B.   C.   D.   E.

3. Which does *not* characterize double-groove welds?
   A. Less distortion of the weldment.
   B. Less metal used.
   C. Higher cost.
   D. 100% penetration.

4. Fillet welds made from each side of the joint may be (*smaller, larger*) than those made from one side of the joint.

5. What type of joint is used in the auto industry for overlapping sheet metal pieces with a spot or plug weld?

6. In what weld configuration are lands used at the bottom of the joint for butt welding pipes together to minimize shrinkage?

7. Prefabricated _____ are used in butt welds to assemble the parts and minimize burn-through.

8. On welding drawings, the symbol below the reference line indicates the weld is to be made on the _____ of the joint. The symbol above the reference line indicates the weld is to be made on the _____ of the joint.

9. Identify the weld symbols shown.

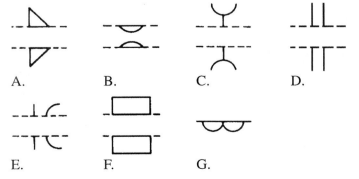

A.   B.   C.   D.   E.   F.   G.

10. Name each welding position described:
    A. Is the most costly position with regard to metal deposition.
    B. In this position, sagging can result when heat rises.
    C. Undercut can be a major problem in this position.
    D. This position has few problems with porosity.
11. How can full penetration (and strength) be achieved with a butt weld?
12. When two pieces of metal do not line up, they have a condition that is called _____.
13. What is transverse shrinkage?
14. Why is less shrinkage involved in double-groove welds than in single-groove welds?
15. Two cylindrical parts which may be fitted together by heating and shrinking produce a(n) _____ type joint.
16. What is the rule of thumb for fillet weld lengths?
17. How can lack of penetration at the joint intersection of fillet welds be prevented?
18. Which type of weld should *not* be used in structural assemblies?
19. Explain how preheating and postheating are used.
20. How is heat removed during the welding operation?

CHAPTER 7

Welding Procedures and Techniques

Objectives

After studying this chapter, you will be able to:
- List welding procedure requirements that must be determined before the actual weld is made.
- Determine when a certain welding technique should be used.
- Recognize a weld schedule.
- Troubleshoot common FCAW problems that occur during initial setup and production.

Important Terms

drag angles
electrode stickout
oscillated beads
pull welding
push welding
stringer bead pattern
triangular weave bead pattern
undercut
visible stickout
weld schedule

Basic Operation

FCAW is either a semiautomatic or fully automatic operation. In semiautomatic welding, technique is critical to the quality of the weld. The welder must be able to set up the machine and operate the gun properly to produce a weld that meets strict fabrication requirements. In a fully automatic operation, welding requirements are controlled by the machine, and operator skill is not important.

Welding Procedure Requirements

The proposed welding procedure should be made and proven to establish the filler materials for the specific base materials, prove the joint design, and obtain the actual welding parameters. The basic procedure should include the areas described next.

Base materials and joint types

Base materials are indicated by specification, grade, type, and thickness. Joint design provides details on the type of joint, including bevel angles, root spacing, type of backing material, and specifications for backgouging groove welds.

Filler materials

Filler materials include electrodes made for out-of-position welds, high-deposition rates for groove- or build-up-type welds, and welding stainless steels, cast irons or various grades of carbon steel. The third number in an AWS specification designates the position for which the electrode is designed (refer to Figure 5-2).

Electrode diameters range from 0.030″ to approximately 0.150″ and are available in fractional and decimal sizes. In general, the smaller electrodes are used for out-of-position welding, and the larger electrodes are used for flat and horizontal welding. For the lowest arc time and cost, use the largest possible electrode size.

When using a constant voltage (CV) power supply and a preset arc voltage, electrode speed will determine the amount of welding current. See **Figure 7-1.** To set the feed rate, run the electrode out of the gun for ten seconds. Measure the length of the electrode and multiply it by 6 to determine the amount of electrode feed for one minute. A switch, rather than a potentiometer, is used to increase or decrease the feed rate for the amount of current desired.

When using a constant current (CC) power supply and a voltage-sensing wire feeder, the welding current and amount of filler material fed into the molten pool will determine the amount of arc voltage. Electrode feed is set using the manufacturer's set-up graph or formula,

Figure 7-2. Electrode feed measured at the tip of the gun will not be accurate since the wire feeder is a voltage-sensing unit and reads a higher OCV from the power supply.

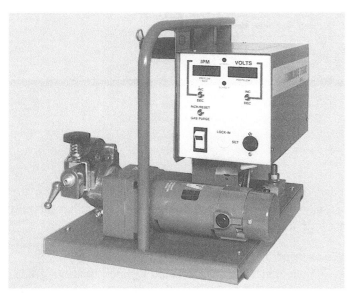

Figure 7-1. Digital CV electrode feeder/control displays electrode speed in inches per minute (ipm). Gas shielding requirements can be set using controls for gas preflow, gas postflow, and purging. (L-TEC)

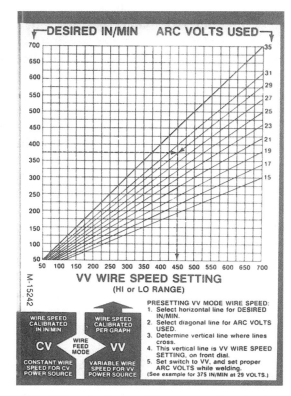

Figure 7-2. IPM and voltage settings on a setup chart are used to establish feeder potentiometer settings on a CC power supply. (Lincoln Electric Co.)

Current type and polarity

Polarity is set on the power supply (if a switch is available), or the electrode leads are connected to the appropriate output terminals. For DCSP (straight polarity), the power cable is connected to the electrode terminal, and the ground cable is connected to the remaining terminal. For DCRP (reverse polarity), the power cable is connected to the ground terminal, and the ground cable is connected to the remaining terminal.

When using a CC wire feeder, the cables must be properly connected to the power supply, and the polarity switch must be in the correct position for the unit to operate. See **Figure 7-3.** These units are designed to shut down if connections are not properly made. The ground cables connect the power supply and the workpiece.

Arc voltage

Arc voltage, which determines the gap between the end of the electrode and the workpiece, is controlled by a potentiometer on the front of a CV power supply, **Figure 7-4.** To increase or decrease the arc gap, turn the dial to a higher or lower number, respectively. It is possible to preset the welding voltage. Turn on the machine, set the voltage dial to midrange, and operate the trigger. Observe the dial; then, set the voltmeter 3 volts higher (for each 100 amperes) than desired. Voltage can be adjusted during welding to increase or decrease the arc gap.

On a CC wire feeder power supply, arc voltage is set by first establishing the desired amperage output. (Use a portable ammeter, if required.) Next, set up the wire feeder according to the manufacturer's set-up graph or formula. Start welding, adjusting the amperage to increase or decrease voltage. The voltage-sensing circuit will control the electrode speed to maintain the welding voltage.

Figure 7-3. The electrode cable from the CC power supply is directly connected to the electrode feed block. (Lincoln Electric Co.)

Figure 7-4. This CV welding power supply has a main voltage rheostat. The machine supplies sufficient current to maintain the preset welding voltage. (Lincoln Electric Co.)

Shielding gas

Shielding gas type, if used, is specified by the electrode manufacturer. Shielding gas is important to the deposition of the molten metal and final metallurgical properties of the deposit. Gas nozzle and gas flow rate must be properly selected. Flow rates range from 35 cfh to 50 cfh when welding in still air. Avoid welding in windy or drafty areas where gas coverage is poor.

Stickout

Stickout is the distance from the end of the contact tip to the end of the electrode. It is also called the contact tip-to-work distance or *electrode stickout* (ESO). Each electrode manufacturer specifies stickout distance, which the welding operator must maintain. See **Figure 7-5.**

When long stickouts (and long nozzles) are used for high-deposition welding, the contact tip is not visible to the operator. *Visible stickout* refers the distance between the end of the gas nozzle and the end of the electrode that *can* be viewed. See **Figure 7-6.** Special nozzles and electrodes are used.

Increasing the stickout will lower the current at the weld pool and lower the voltage across the arc. Lowering arc voltage increases bead convexity (height of crown), decreases porosity, and reduces penetration. Decreasing the stickout increases weld current at the molten pool and increases penetration.

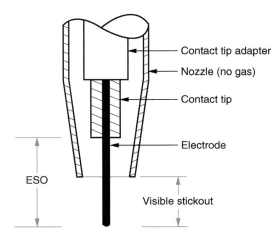

Figure 7-6. The welder cannot see the end of the contact tip. Electrode stickout is measured from the end of the special nozzle to the end of the electrode.

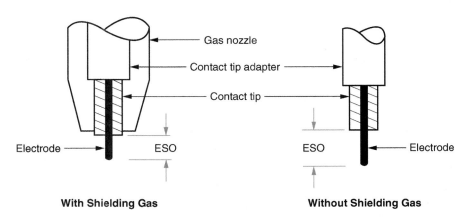

With Shielding Gas   **Without Shielding Gas**

Figure 7-5. Electrode stickout (ESO) is controlled by the welder during the welding operation. To maintain an even flow of current to the electrode, the stickout distance must be held as closely as possible to the manufacturer's recommendation.

Welding Techniques

In an automatic welding operation, travel speed does not vary, as it does in a semiautomatic (manual) operation. In the manual mode, the welder can adjust speed and other variables to produce a satisfactory weld. Practice in making all types of joints, especially those requiring multiple-pass welds, is essential to developing proper technique. Every pass affects the weld. Traveling too slowly produces wide welds, while traveling too quickly reduces width.

The gun may be pointed either toward the molten pool (called *pull welding* or *backhand welding*) or away from the molten pool (called *push welding* or *forehand welding*). **Figure 7-7** shows the backhand technique for welding in the flat and horizontal positions. The forehand gun angle is not often used for welding in the flat position, since slag tends to flow forward and become trapped in the weld. For welding in the vertical position, both the pull and push techniques are used, **Figure 7-8.** Overhead position welding is generally done using the backhand (pull) technique.

Gun angles

Gun angles are either longitudinal (along the weld) or transverse (across the weld), **Figure 7-9.** Angles for backhand welding on linear welds are called *drag angles* and have a considerable effect on penetration, bead form, and final-weld bead appearance. Adjust gun angle and travel speed to maintain good weld pool control and weld bead shape.

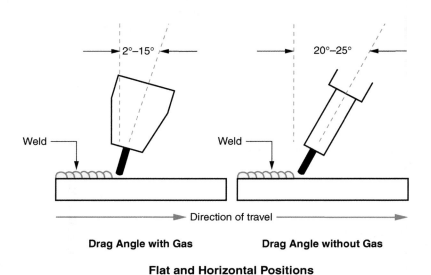

Flat and Horizontal Positions

Figure 7-7. Penetration into the base metal is deeper with a slight gun angle. Weld crown height will increase as the gun angle is increased.

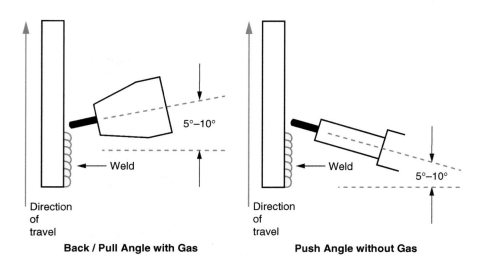

Figure 7-8. Tilting the gun into a pull angle will keep the shielding gas on the molten pool. When welding with a self-shielding electrode in the vertical position, the push technique can be used.

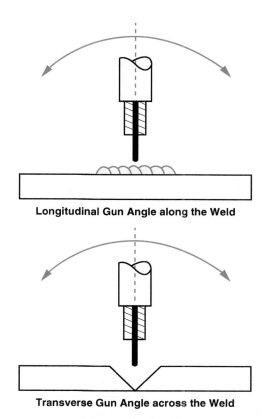

Longitudinal Gun Angle along the Weld

Transverse Gun Angle across the Weld

Figure 7-9. Direction of gun angles for longitudinal and transverse welds.

FCAW electrodes contain different amounts of flux. Therefore, gun angles will vary from slight (for small amounts of flux deposition) to considerable (for electrodes with substantial flux content). If the drag angle is insufficient, the flux will move around to the front of the pool and become entrapped in the weld.

Gun angles for bead placement for flat and horizontal

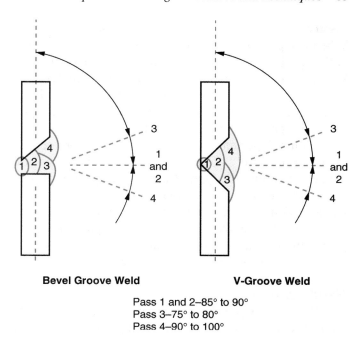

Bevel Groove Weld   **V-Groove Weld**

Pass 1 and 2–85° to 90°
Pass 3–75° to 80°
Pass 4–90° to 100°

Figure 7-11. Horizontal welds tend to fall when too much metal is deposited on the weld pass. Move faster and make stringer beads to control bead formation. Fill the groove from the bottom up.

groove welds are shown in **Figures 7-10** and **7-11.** These angles and bead placement positions may also be used for overhead welding. In the horizontal and overhead positions, weld metal tends to flow downward, causing poor bead formation. Such welds should be made with stringer beads to give the welder better control of the weld metal and final bead surface contour.

Bead sequence

When large fillet welds are required, use multiple-pass welds and start the bead sequence as shown in

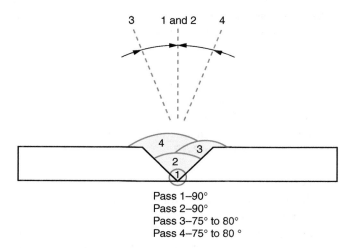

Pass 1–90°
Pass 2–90°
Pass 3–75° to 80°
Pass 4–75° to 80 °

Figure 7-10. This sequence of bead deposition is used for heavy joints. Always deposit weld metal to permit sufficient area for the next pass.

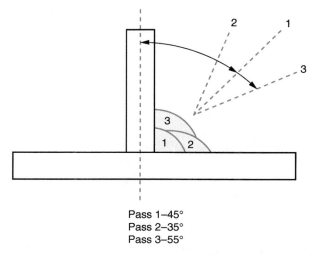

Pass 1–45°
Pass 2–35°
Pass 3–55°

Figure 7-12. Large horizontal fillets require more than one pass. Do not try to make the weld in one or two passes.

Figure 7-12. Work upward, adding passes as required. When using a gas-shielded electrode *uphill,* use a slight pull angle from 0° so the gas will cover the molten pool. When using a self-shielded electrode in the vertical position, use a slight push angle for better slag control and visibility of the molten pool. Refer to Figure 7-8 for gun angles for groove and fillet welding uphill.

Both stringer and wash beads are used for welding uphill. Successful vertical welds depend on several factors, including the type of electrode, operator ability, quality desired in the weld, and final crown shape. When welding *downhill,* the gun angle must be a pull-type leading angle, **Figure 7-13.** Insufficient gun angle will permit the slag to run into the molten pool, resulting in slag pockets or incomplete penetration.

Bead patterns

With a ***stringer bead pattern***, travel is along the joint. The weave is made only enough to prevent slag entrapment in the weld and ensure the molten metal washes into the previous weld or base material. See **Figure 7-14.**

A ***weave bead pattern*** (also referred to as wash beads or ***oscillated beads***) is used to flow metal over a wider area, **Figure 7-15.** Because travel speed is slower, weave beads create more heat than do stringer beads, resulting in greater distortion. A depression (***undercut***) will form at the end of the bead if a dwell (wait) period is not allowed at the end of the pass. See **Figure 7-16.**

A ***triangular weave bead pattern*** is often used in welding the first pass on heavy plate in the vertical

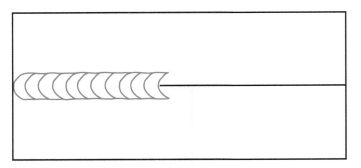

Figure 7-14. A stringer bead pattern is made with little or no side-to-side motion.

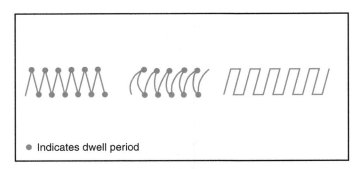

Figure 7-15. Types of weave bead patterns.

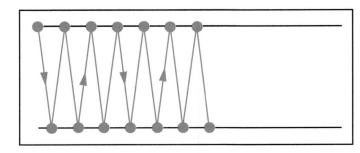

Figure 7-16. Oscillated (wash) beads are made to cover previous welds and require a dwell at the end of the weave to prevent undercut. The amount of forward movement at each dwell can be determined by closely watching the metal fill at the edge of the weld and the height of the new pass or layer.

Figure 7-13. Welding downhill requires high current and fast travel speed to keep the molten flux from moving ahead of the weld pool.

position. Since this technique generates considerable heat, it cannot be used on thin materials. Considerable practice is required to be able to deposit the weld metal in consistent layers with adequate penetration into the fillet corner. See **Figure 7-17.**

Test Welds

The actual welding of component parts results from a combination of parameters and variables (materials, wire sizes, types of gases, welding techniques). The proposed welding procedure must be tested before the actual weld is made. To determine the actual values, select an initial setting for the type and thickness of

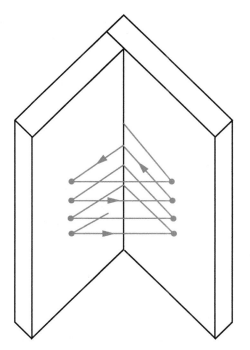

Figure 7-17. Triangular weave patterns require practice to correctly time upward movements and deposit each layer in the proper place. Use this type of weld for root weld passes only.

material involved. (Settings are found in the operating parameter charts in the Reference Section.)

Prepare the joint (if required), clean the material, and make a series of test welds. During the test period, adjust the machine and welding techniques until the desired weld is produced. When the test weld meets all the visual and mechanical requirements of the specification, record the parameters and variables on a *weld schedule,* **Figure 7-18.** Recording the data for a particular weld enables the weld to be duplicated at a future date.

A weld is like a signature—unique to every individual. Each welder has his or her own technique, meaning no two welds are exactly alike. You sign your work, so do your best on any job. Because machine welds can be duplicated, the tolerance for variables can be smaller or tighter.

Operational Problems

During initial setup and production, problems can occur that affect the FCAW welding operation. Many of these problems can be corrected with some simple

---

FLUX CORED ARC WELDING (FCAW) SCHEDULE

WELD TYPE _____ POSITION _____

BASE METAL TYPE _____ THICKNESS _____

ELECTRODE _____ CLASS _____ DIA. _____

ELECTRODE IPM _____ AMPS _____ ELECTRODE STICKOUT _____

CURRENT TYPE _____ VOLTAGE (ARC) _____

GAS TYPE _____ % _____ GAS TYPE _____ % ___ CFH _____

NOZZLE TYPE _____ NOZZLE DIAMETER _____

PREHEAT TEMPERATURE MIN. _____ POSTHEAT TEMPERATURE _____

INTERPASS TEMPERATURE _____ TIME _____

MIN. _____ MAX. _____

NOTES AND SPECIAL INSTRUCTIONS:

Figure 7-18. A welding schedule is a written record of the parameters and variables of a particular weld.

changes, as outlined in **Figure 7-19.** Chapter 8 will describe specific weld defects and corrective steps that can be taken.

Review Questions

Please do not write in this text. Write your answers on a separate sheet of paper.
1. List seven areas that must be determined before the actual weld is made.
2. What determines the amount of welding current on a CV power supply?
3. What determines the amount of arc voltage on a CC power supply?
4. What are the alternate names for the two types of current mentioned in this chapter?
5. Define *electrode stickout* and *visible stickout.*
6. For what type of weld are long stickouts and long gas nozzles used?
7. You are welding with a self-shielded filler material and want to increase penetration into the base material. What do you do?
8. When is the push or forehand welding technique generally used?
9. Why is the push technique *not* often used in the flat position?
10. What technique should be used when welding in the horizontal and overhead positions?
11. What problem will occur if oscillated beads are not properly made?
12. What is the purpose of a welding schedule?

Problem	Possible causes	How to correct
Electrode stubs against work.	Voltage is too low. Electrode feed is too high. Loose ground connection.	Raise (increase) voltage. Decrease electrode feed. Change or tighten ground.
Electrode burns back into contact tip.	Voltage is too high. Electrode feed speed is too low. Stickout is too short.	Decrease voltage. Increase electrode feed speed. Increase stickout.
Electrode feed is erratic.	Kinked electrode feed conduit. Electrode feed drive roller slipping. Kinked filler electrode. Brake adjustment incorrect. Contact tip partially clogged.	Straighten. Adjust idler roll. Check filler electrode for kinks. Check brake tension. Clean ream hole.
Arc blows metal from intended path.	Poor ground location. Insufficient ground clamp area.	Change ground location. Add another ground.
Weld pool extremely fluid.	Current too high.	Reduce electrode speed or amperage.
Weld pool very sluggish.	Current too low. Stickout too long.	Increase electrode speed or amperage. Decrease stickout.
Arc outages.	Contact tip hole oversize. Poor ground connection.	Change contact tip. Change ground clamp.

Figure 7-19. A troubleshooting guide for operational problems during FCAW.

8 Inspection, Defects, and Corrective Action

Objectives

After studying this chapter, you will be able to:
- Define a qualified welding procedure.
- Describe five nondestructive inspection methods.
- Troubleshoot defects for groove, fillet, plug, and spot welds.

Important Terms

defects
destructive testing
discontinuities
liquid penetrant inspection
magnetic particle inspection
nondestructive testing
qualified
radiographic inspection
ultrasonic inspection
visual inspection

Qualified Welding Procedure

Every flux cored arc weldment must meet a standard of quality for the particular weld used. Determining whether a standard of quality has been met can range from a casual look at a weld to see if the parts are joined, to a prescribed inspection for compliance with a specification. High-quality welds may require the use of test welds to establish correct parameters before production begins. Nondestructive testing is performed to verify quality, while destructive testing is performed to verify mechanical values. If the tests are acceptable, the welding procedure is considered *qualified,* and the product can be welded using the procedure. If changes are made to the welding values after qualification, testing must be repeated to requalify the new procedure.

Remember: Any time you set up a machine, change the electrode, install a gas supply, or modify a parameter, run a test weld on a piece of scrap material. It is far better to throw away a piece of scrap metal than to ruin a production part.

Inspection Methods

All welds have flaws, or *discontinuities.* The role of inspection is to locate and determine the extent of those flaws. Extensive flaws are called *defects* and may cause a weld to fail. *Nondestructive testing* of a weld or assembly verifies quality and does not cause damage. The part is usable after the testing is done. *Destructive testing* is performed to determine the physical properties of a weld. Only nondestructive inspection methods are discussed in this chapter and include visual, liquid penetrant, magnetic particle, ultrasonic, and radiographic testing.

Visual Inspection

Visual inspection involves viewing the weld on the front or top surfaces (and penetration side, if visible) plus using rulers, scales, squares, and other tools to determine the condition of the weld. Common defects that can be identified by nondestructive visual examination include:
- Crown height.
- Crown profile.
- Underfill or low weld.
- Undercut.
- Overlap.
- Surface cracks.
- Crater cracks.
- Surface porosity.
- Weld size.
- Weld length.
- Joint mismatch.

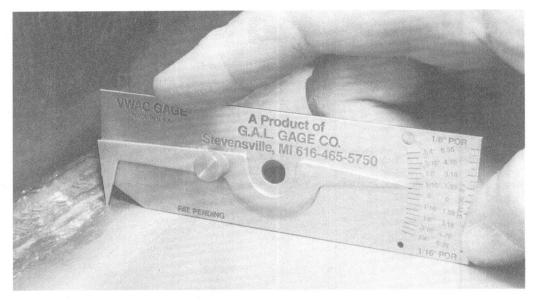

Figure 8-1. An inspection tool is used to check undercut at the edge of a weld. The amount of undercut is shown on the scale. (G.A.L. Gage Co.)

- Warpage.
- Dimensional tolerances.
- Root side penetration.
- Root side profile.

The ripples in the weld should be even, without high and low areas or undercut at the weld toe. A special inspection tool can be used to check undercut depth, crown height on a butt weld, and drop-through on the penetration side of the butt weld. See **Figure 8-1.**

Fillet welds require the fillet leg be a specific size. Unless otherwise specified, the legs should be the same length, **Figure 8-2.** Fillet welds may also require inspection of the throat dimension, **Figure 8-3.**

Liquid Penetrant Inspection

Liquid penetrant inspection is a nondestructive test performed on the surface of a weld or the penetration side of a butt weld. Colored liquid dye and fluorescent penetrant are applied to the weld. The penetrant seeps into any

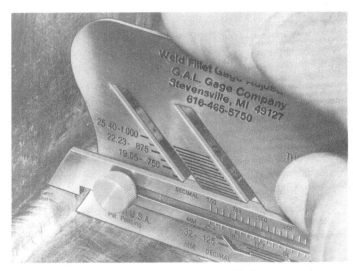

Figure 8-2. The pointer is placed at the edge of the weld leg. Leg size is indicated on the scale. (G.A.L. Gage Co.)

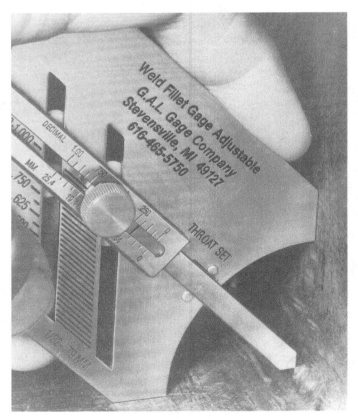

Figure 8-3. Throat size is used to determine concavity or convexity of the weld. (G.A.L. Gage Co.)

cracks or pits; excess penetrant is removed. Liquid developer is applied, drawing some of the penetrant out of the crevices. The dye permits defects to be seen. See **Figure 8-4.** Penetrant inspection does not reveal low welds or undercut. Another variation of this inspection technique uses a black light and fluorescent dye. Penetrant processes can be used in any position on metals, plastics, ceramics, or glass.

Magnetic Particle Inspection

Magnetic particle inspection is a nondestructive method of detecting the presence of cracks, seams, inclusions, segregations, porosity, lack of fusion, and similar discontinuities in magnetic materials. When a magnetic field is established in a ferromagnetic material (iron-based with magnetic properties) that contains one or more defects in the path of the magnetic flux, minute poles are set up at the defects. These poles have a stronger attraction for iron particles than does the surrounding material. The field is magnetized by electric current and a fluid containing iron particles is applied. Any defects are shown by the pattern of the iron particles. See **Figure 8-5.** This process is mainly used for locating

**Step 1
Penetration**

**Step 2
Rinse**

**Step 3
Development**

**Step 4
Inspection**

Figure 8-4. Liquid penetrant test sequence. (Magnaflux Corp.)

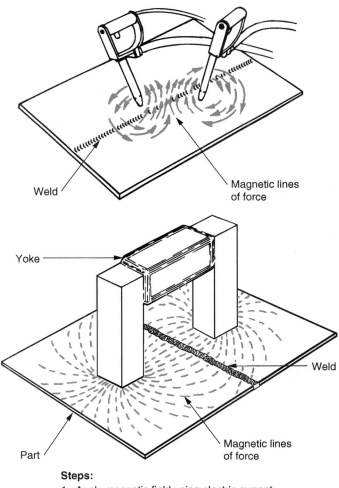

Steps:
1. Apply magnetic field using electric current.
2. Apply magnetic particles while power is on.
3. Blow away excess particles.
4. Inspect.

Figure 8-5. Magnetic particle test sequence.

defects on the surface of the material; however, on thin materials the process is able to locate some defects below the surface.

Ultrasonic Inspection

Ultrasonic inspection is a nondestructive method of detecting the internal presence of cracks, inclusions, porosity, lack of fusion, and similar discontinuities in metals. High-frequency sound waves are transmitted through the part. The sound waves return to the sender and appear on a cathode ray tube (CRT). See **Figure 8-6.** Skilled technicians interpret the test results. Ultrasonic testing has certain advantages, including:

- Superb penetration power, permitting testing of thick materials and a variety of welds.
- Sensitivity sufficient to locate very small defects quickly.
- Ability to be done from one surface.

Radiographic Inspection

Radiographic inspection is another nondestructive test that shows the presence and nature of discontinuities in the interior of a weld. X-rays and gamma rays are used to penetrate the weld. Flaws are revealed on exposed radiographic film, **Figure 8-7.** Although radiographic inspection is expensive compared to other types of tests, the film creates a permanent record of the quality of a weld.

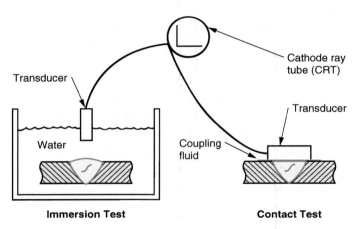

Figure 8-6. During an ultrasonic test, the part and a transducer are submerged in water. When this is not practical, the transducer is coupled (connected) to the test area by a thin layer of liquid.

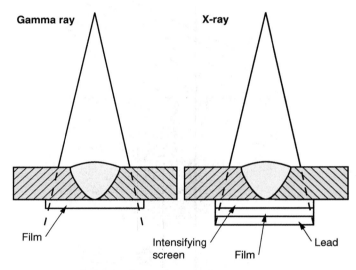

Figure 8-7. Radiographic test operation.

Weld Defects and Corrective Action

Certain defects are common to particular welds. Defects affecting groove, fillet, plug, and spot welds are described next, along with suggestions for correcting each of the problems. More than one corrective action may need to be taken for any given defect.

Groove Weld Defects

Lack of (incomplete) penetration. A weld that does not properly penetrate into the weld joint.

Square groove   V-groove

How to correct:
- Open groove angle.
- Decrease root face.
- Increase root opening.
- Increase amperage.
- Decrease voltage.
- Decrease travel speed.
- Change gun angle.
- Decrease stickout.
- Keep arc on leading edge of molten pool.

Lack of fusion. Fusion did not occur between the weld metal and fusion faces or adjoining weld beads.

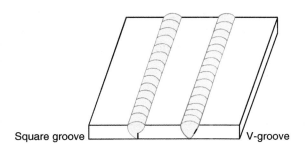

How to correct:
- Clean weld joint before welding.
- Remove oxides from previous welds.
- Open groove angle.
- Decrease root face.
- Increase root opening.
- Increase amperage.
- Decrease voltage.
- Decrease travel speed.
- Change gun angle.
- Decrease stickout.
- Keep arc on leading edge of molten pool.

Overlap. Weld metal that has flowed over the edge of the joint and improperly fused with the parent metal.

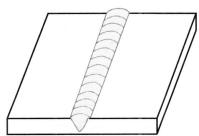

How to correct:
- Clean edge of weld joint.
- Remove oxides from previous welds.
- Reduce size of bead.
- Increase travel speed.

Undercut. Lack of filler material at the toe of the weld metal.

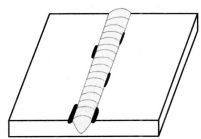

How to correct:
- Decrease travel speed.
- Increase dwell time at edge of joint on wash beads.
- Decrease voltage.
- Decrease amperage.
- Change gun angle.

Convex crown. A weld that is peaked in the center.

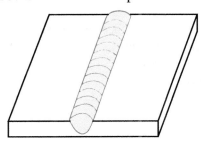

How to correct:
- Change gun angle.
- Decrease voltage.
- Decrease stickout.
- Use wash bead technique with dwell time at edge of joint.

Craters. Formed at the end of a weld bead due to a lack of weld metal fill or weld shrinkage.

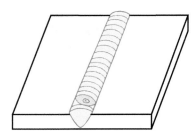

How to correct:
- Do not stop welding at end of joint (use tabs).
- Using same travel gun angles, move gun back on full section of weld before stopping.

Cracks. Caused by cooling stresses in the weld and/or parent metal.

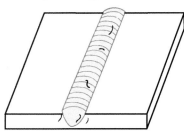

How to correct:
- Use electrode with lower tensile strength or different chemistry.
- Increase joint preheat to slow the weld cooling rate.
- Allow joint to expand and contract during heating and cooling. Increase size of the weld.

Porosity. Caused by entrapped gas that did not have enough time to rise through the melt to the surface.

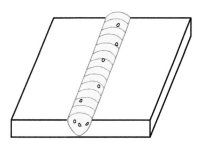

How to correct:
- Remove all heavy rust, paint, oil, or scale on joint before welding.
- Remove flux from previous passes or layers of weld. Increase stickout.

If using shielding gas:
- Check gas flow.
- Protect welding area from wind.
- Remove spatter from interior of gas nozzle.
- Check gas hoses for leaks.
- Check gas supply for contamination.

Linear porosity. Forms in a line along the root of the weld at the center of the joint where penetration is very shallow.

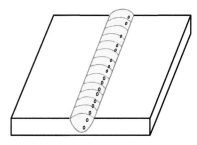

How to correct:
- Make sure root faces are clean.
- Increase current.
- Decrease voltage.
- Decrease stickout.
- Decrease travel speed.

Burn through. Occurs at a gap in the weld joint or place where metal is thin.

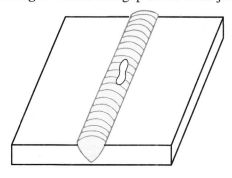

How to correct:
- Decrease current.
- Increase voltage.
- Increase stickout.
- Increase travel speed.
- Decrease root opening.

Excessive penetration. Occurs when the weld metal penetrates beyond the bottom of the normal weld root contour.

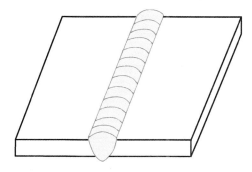

How to correct:
- Decrease root opening.
- Increase root face.
- Increase travel speed.
- Decrease amperage.
- Increase voltage.
- Change gun angle.
- Increase stickout.

Excessive spatter. Forms on the weld and parent metal.

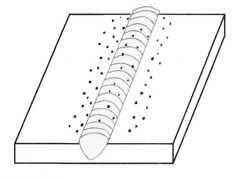

How to correct:
- Decrease voltage.
- Decrease drag angle.
- Decrease travel speed.
- Increase stickout.
- Decrease electrode feed speed.
- Use antispatter spray.

Fillet Weld Defects

Lack of penetration. Insufficient weld metal penetration into the joint intersection.

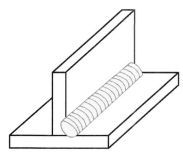

How to correct:
- Decrease gun lead angle.
- Increase amperage.
- Decrease voltage.
- Decrease size of weld deposit.
- Use stringer beads.
- Do not make weave beads on root passes.

Lack of fusion. Occurs in multiple-pass welds where layers do not fuse.

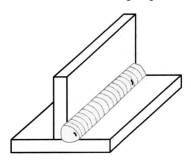

How to correct:
- Remove oxides and scale from previous weld passes.
- Increase amperage.
- Decrease voltage.
- Decrease travel speed.
- Change gun angle.
- Decrease stickout.
- Keep arc on leading edge of molten pool.

Overlap. Occurs in horizontal, multiple-pass fillet welds when too much weld is placed on the bottom layer.

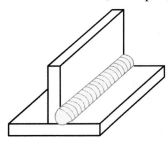

How to correct:
- Reduce the size of the weld pass.
- Reduce amperage.
- Change gun angle.
- Increase travel speed.

Undercut. Occurs at the top of the weld bead in horizontal fillet welds.

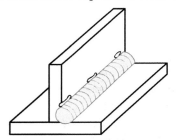

How to correct:
- Make a smaller weld.
- Make a multiple-pass weld.
- Change gun angle.
- Use smaller diameter electrode.
- Decrease amperage.
- Decrease voltage.

Convexity. A weld that has a high crown.

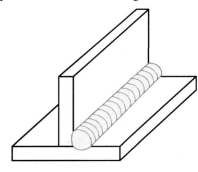

How to correct:
- Reduce amperage.
- Decrease stickout.
- Decrease voltage.
- Decrease gun angle.

Craters. Formed when weld metal shrinks below the full cross section of the weld.

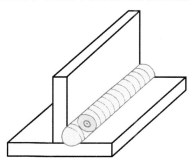

How to correct:
- Do not stop welding at the end of the joint (use tabs).
- Using the same travel gun angle, move the gun back on the full cross section before stopping.

Cracks. Occur in fillet welds just as they do in groove welds.

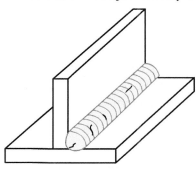

How to correct:
- Use suggestions for groove weld cracks.

Burn through. Occurs when the molten pool melts through the base material and creates a hole.

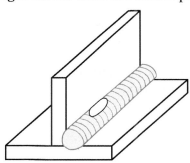

How to correct:
- Decrease amperage.
- Increase travel speed.
- Change gun angle.

Porosity. Caused by entrapped gas that did not have enough time to rise through the melt to the surface.

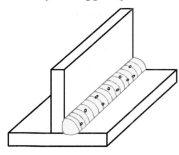

How to correct:
Use suggestions for groove weld porosity.

Linear porosity. Forms along the root of the joint interface.

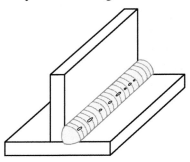

How to correct:
- Use suggestions for groove weld linear porosity.

Plug Weld Defects

Lack of penetration. Occurs when the weld metal does not reach the proper depth in the bottom plate.

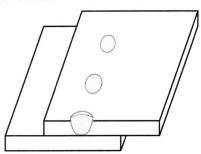

How to correct:
- Increase amperage.
- Decrease voltage.
- Decrease stickout.
- Start weld in center of plug hole, and fill it in a circular pattern.

Excessive penetration. Occurs when the root of the weld extends too far into the bottom sheet of the assembly.

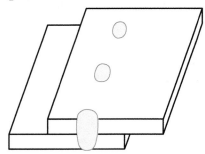

How to correct:
- Decrease amperage.
- Increase voltage.
- Increase stickout.
- Move gun in a circular pattern, and shorten weld operation.

Cracks. Occur at the center of the weld nugget and are caused by rapid cooling of weld metal.

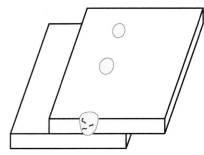

How to correct:
- Use an electrode of lower tensile strength or different chemistry.
- Increase preheat to slow cooling rate.
- Increase size of weld.

Porosity. Caused by entrapped gas that did not have enough time to rise through the melt to the surface.

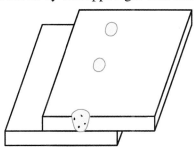

How to correct:
- Use suggestions for groove weld porosity.

Overlap. Occurs at the surface of the weld where too much metal has been deposited.

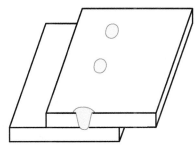

How to correct:
- Shorten welding time so less metal is deposited into the plug weld hole.

Craters. Occur in the center of the plug weld when not enough metal is placed in the hole during the weld cycle.

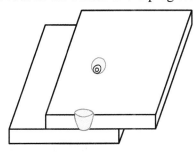

How to correct:
- Lengthen the weld cycle.

Spot Weld Defects

Lack of penetration. Occurs when the weld metal does not reach the proper depth in the bottom plate.

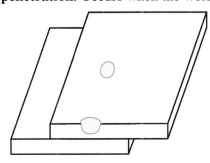

How to correct:
- Increase amperage.
- Increase weld time.
- Decrease voltage.
- Decrease stickout.

Excessive penetration. Occurs when molten metal penetrates through the bottom plate.

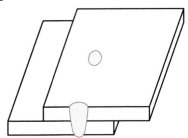

How to correct:
- Decrease amperage.
- Decrease weld time.
- Increase voltage.
- Increase stickout.

Porosity. Caused by entrapped gas that did not have enough time to rise through the melt to the surface.

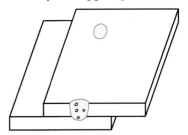

How to correct:
- Increase arc time.
- Increase amperage.
- Decrease voltage.
- Decrease stickout.
- Make sure the metal is clean before mating the assembly.

Cracks. Occur for the same reasons they do in plug welds.

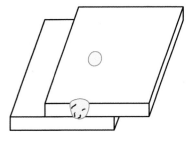

How to correct:
- Use suggestions for plug weld cracks.

Review Questions

Please do not write in this text. Write your answers on a separate sheet of paper.

1. What is a qualified welding procedure?
2. _____ inspection is done by looking at the weld to determine surface irregularities.
3. The tool shown below is used to measure the depth of _____ and to check _____ on butt welds.

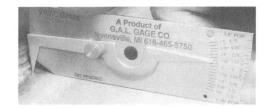

4. Fillet weld inspection includes the size of the _____ and the _____ dimension.
5. _____ inspection uses an electrical current and ferromagnetic material to detect discontinuities.
6. _____ inspection uses high-frequency sound waves for locating discontinuities.
7. Explain how radiographic inspection shows the presence of discontinuities in the interior of a weld.
8. When a weld does not penetrate completely through the base material, the condition is described as _____.
9. A weld that does not have sufficient cross-section at the end of the pass has a _____.
10. _____ is the formation of gas pockets in the weld body or on the crown surface.
11. Molten metal that is expelled from the weld pool is called _____.
12. _____ are caused by the shrinking of the weld and the weld area.
13. _____ affects fillet welds on thin metals and can be caused by welding with excessive amperage.
14. A weld that is peaked in the middle is called a _____.
15. A weld that is caved inward is called a _____.

9 Welding Carbon and Low-Alloy Steels

Objectives

After studying this chapter, you will be able to:
- Describe two methods of metal deposition.
- Select the proper electrode based on usability and performance capabilities.
- Differentiate chemical and mechanical properties for a particular electrode classification.
- Indicate how to safely prepare materials for welding.
- Consider temperature, joint preparation, filler material selection, equipment coordination, proper shielding gases, and test welds when establishing a welding procedure.

Important Terms

carbon steels
ferrous metals
globular arc mode
interpass temperature
postheating
preheating
spray arc mode

Base Materials

The steel family consists of many types and grades. Members of the steel family are considered *ferrous metals* because they contain iron and are magnetic. *Carbon steels* are a group of steels that contain:
- Carbon—1.7% maximum content.
- Manganese—1.65% maximum content.
- Silicon—0.6% maximum content.

Carbon steels are further classified as:
- Low-carbon—up to 0.14% carbon.
- Mild-carbon—0.15% to 0.29% carbon.
- Medium-carbon—0.3% to 0.5% carbon.

- High-carbon—0.5% to 1.7% carbon.

Steels that can be welded by FCAW include:
- Mild and structural steel, identified as ASTM A36, A515, and A516 grades.
- High-strength, low-alloy steels.
- High-strength quenched and tempered alloy steels.
- Carbon-molybdenum steels.
- Chromium-molybdenum steels, such as 1.25% Cr–0.5% Mo and 2.25% Cr–1.0% Mo.
- Nickel steels.
- Manganese-molybdenum steels.

Filler Materials

Filler materials for welding carbon steels are identified in AWS specifications *A5.20 Specification for Carbon Steel Electrodes for Flux Cored Arc Welding* and *A5.29 Specification for Low-Alloy Steel Electrodes for Flux Cored Arc Welding*. Each electrode is identified by a series of letters and numbers that corresponds to the strength, chemical composition, and usability of the electrode. Refer to Figure 5-2 to review how electrodes for carbon and low-alloy steels are identified.

Electrode Characteristics

Each electrode is designed by the manufacturer for a particular type of metal deposition. In the *spray arc mode,* weld metal is deposited as small droplets of metal surrounded by slag and smoke. In contrast, the *globular arc mode* consists of larger globules of metal that detach from the wire in an indefinite pattern. See **Figure 9-1.**

In the electrode classification system, usability and performance capabilities are indicated after the letter T (for tubular electrode). The classification E80T5 Ni$_3$, for example, specifies electrode classification T5. Characteristics of the electrodes available for carbon and low-alloy steels are described next.

Electrode	Characteristics
T1	• Direct current, electrode positive (DCEP). • CO_2 or Ar/CO_2 shielding gas for out-of-position welding. • Single- or multiple-pass welding. • Spray arc mode of transfer with low spatter loss and moderate slag.
T2	• DCEP. • CO_2 shielding gas. • Single-pass welding in flat and horizontal fillet weld positions. • Tolerates some mill scale and rust on material to be welded. • Arc mode similar to T1.
T3	• DCEP. • No shielding gas used. • Spray arc mode of transfer. • Single-pass welds at very high travel speeds in flat, horizontal, and 20° maximum downhill positions on sheet metal up to 3/16″ thick. Not for use on multiple-pass welds.
T4	• DCEP. • No shielding gas used. • Globular arc mode of transfer. • High deposition rate on poor fit ups, with low penetration in the flat and horizontal positions. • Single- or multiple-pass welds.
T5	• DCEP. • CO_2 or 75% Ar/25% CO_2 combination shielding gas. • Globular arc mode of transfer. • Low-hydrogen electrode in same class as E 7018 SMAW electrode. • Excellent crack-resistance and improved impact properties. • Flat and horizontal fillet weld positions with single- or multiple-pass welds.
T6	• DCEP. • No shielding gas used. • Spray arc mode of transfer. • Weld has good impact properties, deep penetration, and excellent slag removal. • Single- and multiple-pass welding in flat and horizontal positions.
T7	• DCEP. • No shielding gas used. • Small-diameter electrodes used for welding in all positions. Large-diameter electrodes restricted to flat and horizontal position welds. • Slag system designed to desulfurize the weld metal, resulting in crack-resistant deposit.
T8	• Direct current, electrode negative (DCEN). • No shielding gas used. • Used for welding in all positions with single- or multiple-pass welds. • Weld is crack-resistant and has good impact properties.
T9	• DCEP. • CO_2 or 75% Ar/25% CO_2 combination shielding gas. • Single- or multiple-pass welding. • Weld is crack-resistant and has good impact properties.
T10	• DCEN. • No shielding gas used. • Single-pass welds on material of any thickness in flat, horizontal, and 20° maximum downhill positions.

Electrode	Characteristics
T11	• DCEN. • No shielding gas used. • Spray arc mode of transfer very smooth. • Single- or multiple-pass welds in all positions. • Thickness limitations. • Preheat, interpass, and postheat temperature requirements for multiple-pass welds.
T12	• DCEP. • CO_2 or 75% Ar/25% CO_2 combination shielding gas. • Single- or multiple-pass welds. • Weld is crack-resistant and has good impact properties.
T13	• DCEN. • No shielding gas used. • Single-pass welding only. • Impact values not specified.
T14	• DCEN. • No shielding gas used. • Single-pass welding only. • Impact values not specified.
TG	See manufacturer's electrode data for information.
TGS	See manufacturer's electrode data for information.

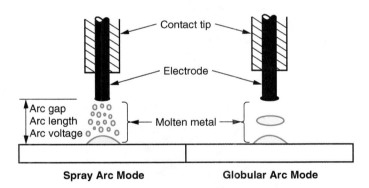

Figure 9-1. In spray arc metal deposition, the weld stream and metal deposition are easy to see. Globular arc metal deposition is distracting to the welder because the metal leaves the electrode tip in a circular pattern.

Chemical and Mechanical Values of Flux Cored Arc Welds

Filler material specifications require welding tests be completed by the manufacturer before acceptance of each lot or heat of flux cored welding electrodes. The tests are performed with qualified welding procedures on a specific type of material. The deposited metal must meet certain chemical, mechanical, and other requirements to meet the specifications. Actual test results of each lot or heat are available to purchasers of electrodes.

Typical results of a chemical analysis for carbon steel deposited weld metal are shown in **Figure 9-2.** Mechanical property test values in the gas-welded condition are shown in **Figure 9-3.**

Typical Electrode Chemical Composition									
	Carbon (C)	Phosphorus (P)	Sulfur (S)	Vanadium (Va)	Silicon (Si)	Nickel (Ni)	Chromium (Cr)	Manganese (Mg)	Molybdenum (Mo)	Aluminum (Al)
T-1 T-4 T-5 T-7 T-8 T-11 T-G	X	0.04	0.03	0.08	0.90	0.50	0.20	0.30	1.75	1.8
X = amount to be determined										

Figure 9-2. Typical chemical analysis for carbon steel deposited weld metal.

Electrode Mechanical Properties			
Electrode Classification	Tensile Strength (ksi)	Yield Strength (ksi)	Percent Elongation 2″
E6XT-1	62	50	22
E6XT-4	62	50	22
E6XT-5	62	50	22
E6XT-6	62	50	22
E6XT-7	62	50	22
E6XT-8	62	50	22
E6XT-11	62	50	22
E6XT-G	62	50	22
E7XT-1	72	60	22
E7XT-2	72	NR	NR
E7XT-3	72	NR	NR
E7XT-4	72	60	22
E7XT-5	72	60	22
E7XT-6	72	60	22
E7XT-7	72	60	22
E7XT-8	72	60	22
E7XT-10	72	NR	NR
E7XT-11	72	60	22
NR = No Requirements			

Figure 9-3. Typical values for electrode mechanical properties for carbon steel deposited weld metal.

Electrode Mechanical Properties			
Electrode Classification	Tensile Strength (ksi)	Yield Strength (ksi)	Percent Elongation 2″
E6XTX-X	6-80	50	22
E7XTX-X	70-90	58	20
E8XTX-X	80-100	68	19
E9XTX-X	90-110	78	17
E10XT-X	100-120	88	16
E11XT-X	110-130	98	15
E12XT-X	120-140	108	14

Figure 9-4. Typical values for electrode mechanical properties for low-alloy steel deposited weld metal.

Electrodes are manufactured for welding several types of alloy steels. Base materials and electrodes designated by major alloy content are as follows:

- *Carbon-molybdenum steel electrodes* have additional molybdenum content.
- *Chromium-molybdenum steel electrodes* have additional chromium and molybdenum content.
- *Nickel steel electrodes* have a higher manganese content and additional nickel, chromium, molybdenum, vanadium, and aluminum.
- *Manganese-molybdenum steel electrodes* have additional molybdenum content.

Proper selection requires matching the type of electrode to the specific base material to obtain the desired chemical and mechanical properties. Mechanical property values of various electrodes used for welding alloy steels are shown in **Figure 9-4.** When selecting electrodes for welding on an alloy steel, all the welding and metallurgical aspects should be evaluated and an appropriate welding test made.

Material Preparation

FCAW can usually be used on slightly rusted material without adverse effect on the completed weld. However, in many cases, specifications do not allow the use of rusted material because of the possibility of defects forming. Plate material should be clean and free of excessive rust, scale, grease, and oil. Weld joint preparation by the thermal-cutting process leaves an oxide film on the surface, where defects can occur, **Figure 9-5.**

Sandblasting will remove scale and rust. Wire brushing or grinding with a power grinder are other methods used, **Figure 9-6.** Wear protective eye and face gear when operating scale or rust removal equipment. When grease burns, it causes gas porosity. Remove grease using a solvent or a degreaser. Do not use solvents or degreasers in a confined space without adequate ventilation or near the arc where the vapors may explode.

Figure 9-5. An oxygen-acetylene cut weld test plate should be smooth, without gouges. The scale on the surface is normal.

Figure 9-6. The weld test plate after a slight face grinding to remove oxidation.

Welding Procedures

The actual joint welding procedure is a combination of joint design and proven welding parameters. Trade journals, welding electrode manufacturers, and others have established various welding procedures. These procedures, and the test results done on similar joints, determine the procedure to use. As you produce the actual joint, consider how you could change various areas to reduce cost or improve the quality of the weld. Major considerations in the establishment of a welding procedure are described next.

Preheat, Interpass, and Postheat

With low-carbon and mild steels, the amount of preheat, interpass temperature, and postheat is not critical unless the material is more than 1″ thick, has a severe joint restraint, or is very cold. These materials do not have sufficient elements to cause hardening of the weld or the weld zone, so cracking is not a problem. As the carbon and alloy content increases, so does the possibility of cracking. Heating temperatures must be considered before, during, and after welding.

Preheating is heating the entire weld area to a specific temperature before welding begins. The temperature should be consistent throughout the thickness of the joint. *Interpass temperature* is the minimum/maximum temperature of the weld metal before the next pass in a multiple-pass weld is made. Interpass temperature should always be maintained until the postheating operation is started. *Postheating* is the final duration and temperature of heating after welding is completed and before the weld is allowed to cool to room temperature.

Several methods can be used to measure temperature, including pellets, crayons, and paints, **Figure 9-7.** The easiest method is to use a crayon or pellet that melts at the desired temperature.

Actual preheat and interpass temperatures used for welding carbon and low-alloy steels are shown in **Figure 9-8.** When welding with alloy electrodes or the higher carbon or alloy steels, consult the manufacturer of the material or the electrode for these temperatures. See the Reference Section for additional temperatures.

Joint Preparation

Joint designs for FCAW follow the same basic design as those for shielded metal arc welding. Gas-shielded and self-shielded operations can both be used, **Figure 9-9.** However, gas-shielded electrodes obtain better penetration with higher current densities. Groove angles may be smaller or the groove joint may have a narrower root opening.

When designing the joint, always consider the movement of the molten metal without undercut and

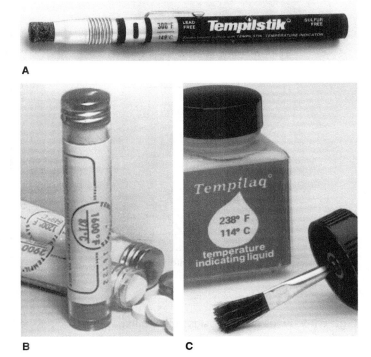

Figure 9-7. These temperature indicators can be applied to the part to be heated and will melt when the specified temperature is reached. A—Crayon. B—Pellets. C—Temperature indicating liquid.

Plate Thickness in. (mm)	Up to ¾ (19)	¾-1½ (19-38)	1½-2½ (38-64)	Over 2½ (64)
Recommended minimum preheat temperature, °F (°C)	70 (21)	150 (66)	150 (66)	225 (107)
Recommended interpass temperature, °F (°C)	70 (21)	150 (66)	225 (107)	300 (149)

Figure 9-8. Recommended initial temperatures for preheat and interpass. Higher temperatures may be required depending on job conditions, codes, or the presence of cracks.

visual access of the joint by the welder. The final joint design should be verified by a test to establish the welding parameters for actual production welding.

When making fillet welds using gas shielding, the fillet weld can be made smaller but still have the same strength as a self-shielded weld. Cost savings can be considerable if this is done. See **Figure 9-10.**

Filler Material Selection

To select the proper filler material, review the joint for the following factors:

- Select an electrode with sufficient mechanical and physical properties for the base material to be welded.

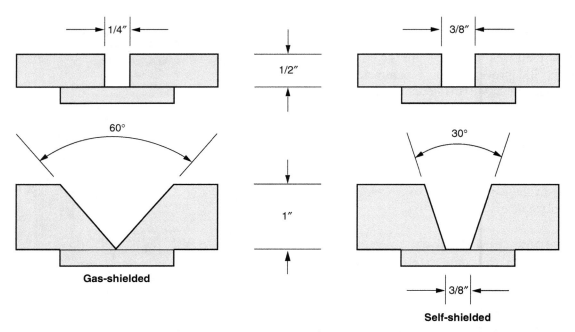

Figure 9-9. Types of joint designs used with gas-shielded and self-shielded FCAW electrodes. Gas-shielded electrodes have greater depth of penetration.

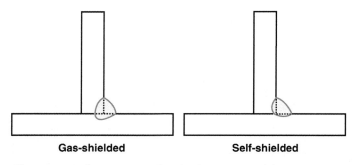

Figure 9-10. Gas-shielded fillet welds in the flat and horizontal positions can be made smaller than the self-shielded type since penetration into the corner is deeper.

- Select a type and diameter of electrode for the joint involved.
- Select an electrode for the position in which the welding will be performed.
- Select an electrode for either single- or multiple-pass welding.
- Determine if the electrode will be gas-shielded or self-shielded.
- Review deposition rates to determine the cost between various types and sizes of electrodes.

Several areas of equipment coordination must be considered. They include:

- Proper type of welding equipment for the intended electrode.
- Welding machine amperage capacity, including duty cycle.
- Type of gun, liners, contact tips, and nozzles.
- Wire feeder type, drive roller configuration, and size.

Shielding Gas

The manufacturer specifies the proper shielding gas for a particular electrode. Always use the recommended gas or gas mixture unless you have specific instructions to do otherwise. Sufficient gas flow over the molten pool is required to prevent air from entering and to enable the arc to melt off the wire in the proper pattern. Wind drafts disturb normal gas flow and may create problems with the arc pattern, leading to porosity in the weld.

Gas nozzles should be large enough to cover the molten weld pool and distribute the flow of gas over the entire weld width. During welding, spatter should be removed from the gas nozzle to allow proper gas flow. Antispatter compounds are available to prevent the buildup of spatter in and on the nozzle tip, **Figure 9-11.**

Test Welds

Established welding procedures for the carbon steels are given in Chapter 14 and the Reference Section of this textbook. Wire type and size, amperage, polarity, stickout, position of welding, joint design, and other areas must be considered before starting the actual welding operation. Pay attention to the following factors:

- Welds should be free of visible defects, such as cracks, porosity, and undercut. If defects are present, the welding technique or procedure must be changed before proceeding.

Figure 9-11. Antispatter spray and gelatin compounds can be applied to the torch nozzle and workpiece any time during the welding operation. Gel should be applied when the nozzle is warm so it melts and flows over the end of the nozzle.

- When making multiple-pass welds, always chip slag from the weld, and clean any residue from the joint with a wire brush.
- Establish gun angles during the test to ensure penetration and bead contour.
- Stringer beads require less skill on the part of the welder; wash beads create more distortion of the weldment.

- To control amperage and voltage, maintain an even electrode stickout during the actual welding operation. Doing so will provide even penetration and proper metal deposition.
- When using a gas-shielded electrode, maintain the gun angle for even gas flow over the molten metal. When using CO_2 as the shielding gas, make sure the regulator does not freeze up and restrict gas flow. With high flow rates, use an electrical regulator with a heater system to avoid freeze-up.
- During welding, check the wire feeder for proper operation.
- If the power supply is used for a long period of time, make sure it is within duty-cycle limits.
- Contact tips are one of the most critical parts in the system. If the tip is worn, the electrical current cannot properly contact the electrode, and an interruption in amperage flow will occur, causing defective welds. Check the contact tip often and replace it when required.
- Gas nozzles will attract spatter during welding. Use an antispatter compound, and clean the nozzle often.

If used properly, test welds can prevent many problems. Not only do they verify the quality of the weld, they serve as a trial-and-error period for the welder to fine-tune the welding procedure, adapt new techniques for handling the gun, and observe the arc and bead pattern. Always record the welding parameters used on the test joint. The recorded procedure can then confidently be used on the actual weld. If changes are made to the procedure, another test joint must be made to verify quality and other required values.

In addition to serving as training and an opportunity to develop accurate procedures, weld tests help lower fabrication costs and encourage shop safety. Equipment condition is another aspect of the test-weld period. Welding equipment and related apparatus, such as cables and ground connection, should be checked during a weld test for proper operation. Know the condition of your equipment, and keep it in good repair.

Defects can and do occur in the welding of carbon and low-alloy steels. Refer to Chapter 8 for a discussion of welding defects and corrective action.

Review Questions

Please do not write in this text. Write your answers on a separate sheet of paper.
1. What are ferromagnetic metals?
2. Describe the two modes of metal deposition for carbon and low-alloy steels.
3. What do the letters DCEP stand for?

Refer to Figures 9-2 and 9-3 to answer the next two questions pertaining to chemical and mechanical properties of electrodes:

4. What is the percentage of molybdenum content in electrode classification T-7?
5. What is the tensile strength of electrode classification E6XT-7?
6. Why must grease and oil be removed from the welding surface prior to welding?
7. What personal protective equipment should be worn when power brushing either the metal or the weld?
8. Name the three types of heating that can be used during the welding procedure.
9. Joint design should always be confirmed by a(n) _____.

10. *True or false?* Fillet welds can be made smaller with the FCAW gas-shielded mode than with the open-arc mode of operation.
11. What are two important considerations concerning the power supply that must be coordinated with the selection of filler material?
12. One of the major areas the welder controls is electrode stickout. Changing this dimension will affect the amperage and voltage of the welding procedure, in turn affecting _____ and proper _____.
13. What can happen to CO_2 regulators if the flow rate is high?
14. How can the condition in Question 13 be prevented?
15. List four advantages of performing a weld test.

10

Welding Chromium-Nickel Stainless Steels

Objectives

- Select the proper electrode based on usability and performance capabilities.
- Indicate how to safely prepare materials for welding.
- Consider joint preparation, filler material selection, equipment coordination, proper shielding gases, tooling, and test welds when establishing a welding procedure.

Important Terms

arc blow
austenitic stainless steels
carbide precipitation
run-off tabs
run-on tabs
stainless steels

Scope

Welding stainless steels with the FCAW process is rather new in the fabrication field. It is restricted to a small number of filler materials for chromium-nickel steels. Other types of electrodes are available for stainless steel alloys used for overlays and for joining dissimilar metals. These electrodes are discussed in Chapter 12.

Base Materials

Chromium-nickel steels are called *stainless steels* because they resist almost all forms of rust and corrosion. They play an important role in the food, drug, and chemical refining industries. Stainless steels are utilized in conditions of extreme heat or cold. They cannot be hardened by heat treatment. *Austenitic stainless steels* contain at least 11% chromium with varying amounts of nickel. Their grain structure is nonmagnetic. **Figure 10-1** lists common base materials in the chromium-nickel group of stainless steels.

Commercially Wrought Stainless Steel Identification (AISI)								
Composition Percent[a]								
Type	Carbon (C)	Manganese (Mn)	Silicon (Si)	Chromium (Cr)	Nickel (Ni)	Phosphorus (P)	Sulfur (S)	Others
201	0.15	5.5– 7.5	1.00	16.0–18.0	3.5– 5.5	0.06	0.03	0.25 N
202	0.15	7.5–10.0	1.00	17.0–19.0	4.0– 6.0	0.06	0.03	0.25 N
301	0.15	2.00	1.00	16.0–18.0	6.0– 8.0	0.045	0.03	
302	0.15	2.00	1.00	17.0–19.0	8.0–10.0	0.045	0.03	
302B	0.15	2.00	2.0–3.0	17.0–19.0	8.0–10.0	0.045	0.03	
303	0.15	2.00	1.00	17.0–19.0	8.0–10.0	0.20	0.15 min.	0–0.6 Mo
303Se	0.15	2.00	1.00	17.0–19.0	8.0–10.0	0.20	0.06	0.15 Se min.
304	0.08	2.00	1.00	18.0–20.0	8.0–10.5	0.045	0.03	
a. Single values are maximum unless indicated otherwise. b. (Cb + Ta) min – 10 × %C. c. Ta – 0.10% max.								

Figure 10-1. American Iron and Steel Institute (AISI) identifies stainless steel by a numbering system and specifies the metal composition of each type.

(Continued)

Commercially Wrought Stainless Steel Identification (AISI)								
Composition Percent[a]								
Type	Carbon (C)	Manganese (Mn)	Silicon (Si)	Chromium (Cr)	Nickel (Ni)	Phosphorus (P)	Sulfur (S)	Others
304L	0.03	2.00	1.00	18.0–20.0	8.0–12.0	0.045	0.03	
305	0.12	2.00	1.00	17.0–19.0	10.5–13.0	0.045	0.03	
308	0.08	2.00	1.00	19.0–21.0	10.0–12.0	0.045	0.03	
309	0.20	2.00	1.00	22.0–24.0	12.0–15.0	0.045	0.03	
309S	0.08	2.00	1.00	22.0–24.0	12.0–15.0	0.045	0.03	
310	0.25	2.00	1.50	24.0–26.0	19.0–22.0	0.045	0.03	
310S	0.08	2.00	1.50	24.0–26.0	19.0–22.0	0.045	0.03	
314	0.25	2.00	1.50–3.0	23.0–26.0	19.0–22.0	0.045	0.03	
316	0.08	2.00	1.00	16.0–18.0	10.0–14.0	0.045	0.03	2.0–3.0 Mo
316L	0.03	2.00	1.00	16.0–18.0	10.0–14.0	0.045	0.03	2.0–3.0 Mo
317	0.08	2.00	1.00	18.0–20.0	11.0–15.0	0.045	0.03	3.0–4.0 Mo
317L	0.03	2.00	1.00	18.0–20.0	11.0–15.0	0.045	0.03	3.0–4.0 Mo
321	0.08	2.00	1.00	17.0–19.0	9.0–12.0	0.045	0.03	5 × %C Ti min.
329	0.10	2.00	1.00	25.0–30.0	3.0– 6.0	0.045	0.03	1.0–2.0 Mo
330	0.08	2.00	0.75–1.5	17.0–20.0	34.0–37.0	0.04	0.03	
347	0.08	2.00	1.00	17.0–19.0	9.0–13.0	0.045	0.03	C
348	0.08	2.00	1.00	17.0–19.0	9.0–13.0	0.045	0.03	0.2 Cu[b, c]
384	0.08	2.00	1.00	15.0–17.0	17.0–19.0	0.045	0.03	
a. Single values are maximum unless indicated otherwise. b. (Cb + Ta) min – 10 × %C. c. Ta – 0.10% max.								

Figure 10-1. *(Continued)*

Filler Materials

Electrodes for chromium-nickel stainless steels are listed in *AWS A5.22, Specification for Stainless Steel Electrodes for FCAW and Stainless Steel Flux Cored Rods for GTAW.* The electrodes are classified by material type rather than by tensile strength (as are mild- and low-alloy steel FCAW electrodes). See **Figure 10-2.** When the word "tensile" appears in the specification, it refers to one of the mechanical properties of the material.

Electrode Characteristics

Electrodes are manufactured by various companies. Not every manufacturer makes every electrode. **Figure 10-3** lists types of stainless steel electrodes currently available for FCAW. **Figure 10-4** shows the operating data for three classes of electrodes. Characteristics of the electrodes available for chromium-nickel steels are described next.

Stainless Steel Electrode Classification	
EXXXT-X	
E	Indicates an electrode.
XXX	Designates classification according to composition.
T	Designates a flux cored electrode.
X	Designates the external shielding medium to be employed during welding.

Figure 10-2. Stainless steel FCAW electrode classification.

Stainless Steel Electrode Types	
E308LT-1	
E309LT-1	(Low-carbon material)
E310T-1	
E312T-1	
E316LT-1	(Low-carbon material)
E317LT-1	(Low-carbon material)
E347T-1	

Figure 10-3. All FCAW stainless steel electrodes operate on DCEP. Most use a 75% AR/25% CO_2 shielding gas mixture, with slightly higher chemical and physical properties.

Chromium-Nickel FCAW Electrode Data		
Electrode Designations	Shielding Gas Types	Current Type/Polarity
EXXXT-1	CO_2	DCEP
EXXXT-2	AR + 2% O	DCEP
EXXXT-3	None	DCEP
EXXXT-G	Not specified	Not specified
Note: Shielding medium should be as specified by the electrode manufacturer, or as defined by chemical and physical results obtained from a qualified welding test.		

Figure 10-4. Chromium-nickel stainless steel electrode operating data.

Electrodes (Class 1)	Characteristics
E308LT-1	• Direct current, electrode positive (DCEP). • CO_2 or Ar/CO_2 combination shielding gas. • For stainless steel (S/S) types 301, 302, 304, 304L, 321 and 347.
E309LT-1	• DCEP. • CO_2 or Ar/CO_2 combination shielding gas. • For S/S Type 309 in cast or wrought form and Type 304 to mild steel.
E310T-1	• DCEP. • CO_2 or Ar/CO_2 combination shielding gas. • For S/S Type 310 in cast or wrought form and dissimilar metals.
E312T-1	• DCEP. • CO_2 or Ar/CO_2 combination shielding gas. • For joining dissimilar steels.
E316LT-1	• DCEP. • CO_2 or Ar/CO_2 combination shielding gas. • For S/S Type 316 with low carbon to prevent carbide precipitation and intergranular corrosion.
E317LT-1	• DCEP. • CO_2 or Ar/CO_2 combination shielding gas. • For stainless steels with molybdenum for corrosion resistance.
E347T-1	• DCEP. • CO_2 or Ar/CO_2 combination shielding gas. • For columbium stabilized S/S Types 321 and 347.

Electrodes (Class 2)	Characteristics
EXXXT-2	• Ar/2% O_2 combination shielding gas that provides a definite spray-arc pattern of the molten metal. • No electrodes currently available in Class 2 category.

Electrodes (Class 3)	Characteristics
EXXXT-3	• Shielding gas not required. • Direct current, reverse polarity (DCRP) in all positions. • Several electrodes available in Class 3 category.

Class 1 and Class 3 electrodes deposit metal with the same basic chemistry. Class 1 electrodes transfer more of the required elements as the shielding gas provides a pathway for the molten metal and prevents air from entering the welding area. If chemical or physical tests are required, weld tests must be completed to determine the proper electrode, shielding gas, and welding procedure.

Chemical and Mechanical Values of FCAW Electrodes

The American Welding Society requires tests on specific base materials for each type and lot of electrode. The deposited metal is then tested and must meet specific chemical and mechanical requirements to determine compliance with specification *AWS A5.22*. Chemical and physical test reports are available from electrode manufacturers for every heat or lot of electrode. Individual spools or coils of electrode have tags identifying the heat or lot.

Material Preparation

Stainless steel has a corrosion-resistant film on its surface caused by a reaction between the chromium/nickel and the atmosphere. For the most part, the film does not interfere with the quality of the completed weld. If any corrosion *is* present, it should be removed by grinding or brushing before welding.

Caution: Do not use carbon steel wire brushes on the parent metal or the weld. Always use stainless steel material brushes. If carbon steel brushes are used on the parent metal before welding, carbon will be absorbed into the molten metal and could lead to carbide precipitation and intergranular corrosion.

Joint preparation done by plasma-arc cutting should be ground to bright metal before welding. Otherwise, oxides from the cutting will be absorbed into the weld and cause defects. Oil or grease on the surface of the material should be removed before welding, as well.

Welding Procedures

Welding procedures for stainless steels are almost the same as for carbon steels. The one major difference is the diameter of the electrode used for welding. Where carbide precipitation is a problem, use small-diameter electrodes to keep down heat input. This is also true when welding in the overhead position where large molten pools cause metal sagging and excessive crown height.

Use **run-on tabs** and **run-off tabs** on longseam welds to obtain penetration at the start of the weld and to prevent craters at the end of the weld. See **Figure 10-5.** Craters do not have enough weld cross-section, so cracks will form in these areas. A small tack weld at the longseam intersection will keep the tabs in place during the welding operation. If tabs cannot be used, reverse the direction of the arc when you stop welding to build up metal and prevent craters.

Stainless steels retain heat much longer than carbon steels, so tooling may be necessary to remove heat from the weld area immediately after welding. Do not use carbon-steel tooling; carbon will transfer to the stainless steel wherever contact is made between the two.

Remember: Stainless steel does not typically rust. Rust on stainless steel is caused by contact with carbon steel. Rust must be removed from carbon steels with a cleaner or acid to prevent transference to stainless steels. Carbon steel is magnetic. When welding carbon steel to stainless steel, arc blow may result if the ground is not properly placed.

Figure 10-5. The arc is started on one end of the tab, and the weld continues to the end of the other tab. Good penetration at the start and stop of the weld, without cold spots and craters, is ensured.

Preheat, Interpass, and Postheat

Chromium-nickel steels generally do not require preheat, interpass, or postheat cycles. Problems are likely to occur when the base material gets hot. Stainless steels retain heat during welding, leading to cracks and *carbide precipitation,* a condition in which chromium leaves the grains and combines with carbon in the grain boundary. If acid were to enter the grain boundary, it could cause intergranular corrosion or weld failure. To prevent cracking, use a low- or extra-low-carbon filler material. Keep the weld heat to a minimum. Use stringer beads at a maximum interpass temperature of 300°F (149°C). The material should not be welded below 60°F (16°C). These types of stainless steels do not require postheating.

Joint Preparation

Joints for chromium-nickel stainless steels are prepared similarly to joints for mild-steel materials, with the following exceptions:

- Self-shielded electrodes produce a slightly sluggish weld that may require more groove width and a larger bevel angle for manipulation of the molten metal. Joint groove widths and bevel angles should be fit up on the high side of the joint tolerance.
- Gas-shielded electrodes require a gas nozzle on the end of the welding gun. Joint opening should be sufficient to make the joint and electrode extension visible so the operator can manipulate the gun.

Filler Material Selection

Selection of the proper filler metal involves reviewing the joint for the following factors:

- The electrode should closely match the composition of the base material. Always use a low-carbon electrode for low-carbon material. When mechanical property requirements are imposed, check the test report for verification.

- Select an electrode diameter for the type of joint to be welded.
- Select an electrode type and diameter for the position in which the welding will be performed.
- Determine whether the electrode will be self-shielded or gas-shielded.
- Review the deposition rates to compare the deposition costs of various sizes of electrodes.

Once the filler material is selected, welding equipment must be coordinated. Areas of consideration include:

- Machine capacity.
- Duty cycle.
- Type of gun and nozzles.
- Contact tips.
- Cable liners. Always use nylon or Teflon™ liners with stainless steel electrodes. Check them often during the operation for replacement.
- Wire feeder drive roll size and design. Drive rolls may be either V- or U-type; however, use only finger-tight feed-roll pressure. Too much pressure will cause the electrode to crack and the liners to wear excessively.
- Type of shielding gas.

Shielding Gas

The proper shielding gas for a particular electrode is specified by the electrode manufacturer (refer to Figure 10-4). Using a mixture other than the one specified could change the chemical composition of the weld. When chemical requirements are imposed, perform a weld test to verify the final chemical composition complies with the specification. **Caution:** Never use shielding gas on an electrode designated as self-shielded unless gas use is approved by the manufacturer.

Sufficient gas flow is required over the molten pool to prevent air from entering the weld. Wind drafts will disturb normal gas flow and create arc-pattern problems, causing porosity in the weld. Flow rates for gas-shielded welds vary from 25 cfh to 35 cfh, depending on wind gusts in the weld area.

Gas nozzles should be large enough to distribute the shielding gas over the molten pool. Any spatter that accumulates in the nozzle should be removed and an anti-spatter compound applied to the interior of the nozzle.

Tooling and Backing

Nonmagnetic tooling must be used to prevent magnetic *arc blow,* deflection of the intended arc pattern by magnetic fields. This type of tooling also prevents iron pickup on the top and back of the base material. Gas backing can be used to prevent oxidation on full-penetration welds, **Figure 10-6.**

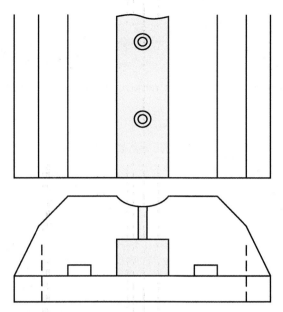

Figure 10-6. Backing bar design for admitting gas to the penetration side of the weld joint.

Where gas backing cannot be used, a flux substance can substitute, **Figure 10-7.** The paste is mixed with alcohol, applied to the part, and allowed to dry before welding, **Figure 10-8.** After welding, the flux is removed with warm water and a stiff stainless steel brush. Further cleaning of the stainless steel may be done with a commercial cleaner, **Figure 10-9.**

Test Welds

As with carbon and low-alloy steels, test welds performed on stainless steels prevent many problems. They verify the quality of the weld and serve as a learning period for the welder. Ultimately, test welds

Figure 10-7. Commercial flux is used when backing bars or tooling cannot be used. (Golden Empire Corp.)

Figure 10-8. Flux powder remains on the part after the fluid evaporates.

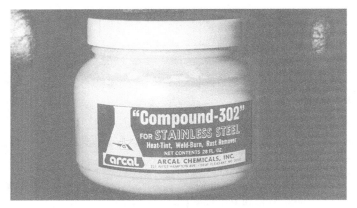

Figure 10-9. Commercial stainless steel weld cleaner. (Arcal Chemical, Inc.)

confirm that the welding procedure is satisfactory and the weld can be properly made. See **Figure 10-10.** The welder has the opportunity to fine-tune the welding procedure, adapt new techniques for handling the gun, and observe the arc and bead pattern. Welding equipment should be checked for proper operation during this period. Check contact tips often for signs of wear.

The Reference Section contains several welding procedures for stainless steels using different types and sizes of electrodes. These procedures are general and should be verified before being used for production.

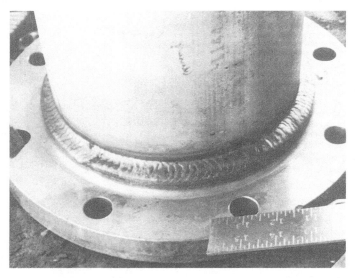

Figure 10-10. Weld tests confirm the welding procedure for this particular weld joint is satisfactory. Destructive tests will ensure the quality of the completed weld.

Problem Areas

Problems can and do occur in the welding of the stainless steels. (Refer to Chapter 8 for a discussion of welding defects and corrective action.) Many specifications and standards for welding require special tests for weld qualification. Others require chemical analysis of the actual deposited weld metal to ensure quality. These measures assure the purchaser the weldment will operate satisfactorily in the environment for which it was made. If the requirements for a particular procedure call for special tests or chemical analysis, investigate the welding areas with a qualified metallurgist before ordering welding material and consumables, designing weld joints, and developing welding procedures.

Review Questions

Please do not write in this text. Write your answers on a separate sheet of paper.

1. Why are chromium-nickel steels called stainless steels?
2. What distinguishes austenitic stainless steels from other stainless steels?
3. How are stainless steel electrodes classified differently than carbon and low-alloy steels?
4. When the word "tensile" appears in the specification for chromium-nickel stainless steel electrodes, what does it refer to?
5. How many classes of stainless steel FCAW filler materials currently exist?
6. What effect does air have on the chromium and nickel content in stainless steels?
7. Explain why carbon steel wire brushes should not be used to remove corrosion from stainless steels.
8. When stainless steels are plasma-arc cut, the cut surface is left with a severe oxide film. How can you prevent oxides from being absorbed into the weld?
9. _____ used at the start and end of a longseam weld will assist in the penetration at the start of the weld and prevent _____ at the end of the weld.
10. Any tooling required for the holding of the base material during welding should be made of _____.
11. Excess heating of the weld joint will cause chromium and carbon to move from the metal grains to the grain boundary. This is called _____ and is one of the primary reasons for cracking in the stainless steels.
12. List three techniques used to prevent cracking in the welds.
13. A groove weld made with a self-shielded electrode requires a(n) _____ for the molten metal to flow properly.
14. Whenever a butt or groove weld is made with full penetration, a gas or _____ backing should be used to prevent oxidation.
15. What does the test weld ultimately do?

CHAPTER 11

Welding Cast Irons

Objectives

After studying this chapter, you will be able to:
- Describe the four types of cast iron.
- Select the proper electrode based on usability and performance capabilities.
- Indicate how to safely prepare materials for welding.
- Consider temperature, joint preparation, filler material selection, equipment coordination, proper shielding gases, and test welds when establishing a welding procedure.

Important Term

ductility

Base Materials

Cast irons contain 93% to 95% iron, 2.5% to 3.5% carbon, and smaller quantities of other elements, such as chromium, nickel, manganese, and copper. The elements vary depending on the type of casting, heat treatment, and final use. Basic composition requirements for cast iron materials are:
- Iron, 93% – 95%.
- Carbon, 1.75% – 4.5%.
- Silicon, 0.5% – 3.0%.
- Manganese, 0.5% – 1.0%.
- Phosphorus, 0.15% – 1.0%.
- Sulfur, 0.05% or more.

As the name cast iron implies, the material is melted and cast (poured into a mold) to shape the desired part. The carbon gives cast iron very high strength and very low ductility. During the initial cooling and solidification, the carbon takes on new forms that determine the final characteristics of the material.

Types of Cast Irons

There are four types of cast irons:
- Gray iron.
- White iron.
- Malleable iron.
- Ductile or nodular iron.

Gray iron is the most commonly used form of cast iron. It casts easily in sand molds and is inexpensive to produce. Heat treatment can be done after casting to improve the mechanical properties. When gray cast iron is cooled, the carbon forms flakes of graphite. The edges of the graphite allow cracks to form when the material is under stress. A cracked surface appears gray, hence the name gray iron. The material has very low *ductility* (ability to deform or exhibit plasticity without breaking). Some alloys of this type of cast iron are readily welded with the FCAW process using specific welding procedures.

White iron is rapidly quenched in a water-cooled mold during casting to prevent formation of graphite in the grain structure. However, graphite does form on the casting surface, which becomes hard and abrasive. The basic casting has very low ductility; therefore, white iron is not considered weldable.

Malleable iron is made by heat treating white iron to change the graphite flakes to a spheroidal shape. This change reduces the tensile strength of the material but improves the ductility. Some of these alloys are readily weldable with the FCAW process.

Ductile iron is also called nodular or spheroidal iron. During the initial melting of the ore, various elements are added to create spheres of the excess carbon, resulting in an extremely strong and ductile material. Various tensile strengths can be made by adding alloys and varying the heat-treating cycles. Higher tensile strengths are directly affected by welding, which may result in the formation of cracks and hard heat-affected zones. All grades of ductile cast iron are weldable using FCAW with the proper welding procedure.

Filler Materials

Filler materials for welding cast irons include steel cored electrodes or an iron-nickel electrode developed for cast-iron applications. The steel electrodes meet specification AWS A5.20 for classifications EXXT-4,5,7, and 8.

The iron-nickel cored electrode is not listed under a FCAW electrode specification. However, the filler material requirements meet specification AWS A5.15 for classification ENiFe-C1.

Electrode Characteristics

The steel electrodes mentioned above do not match any of the chemical requirements of cast iron. Therefore, their use is limited to welding cast iron to steel or to making noncritical repairs.

Because of its mechanical properties, the iron-nickel electrode is an exceptional choice for welding cast iron to cast iron, joining cast iron to other types of metals, and repairing cast iron work.

Characteristics of the electrodes available for carbon and low-alloy steels are described in the chart below.

Chemical and Mechanical Values of Flux Cored Arc Welds

Carbon steel electrodes pick up carbon from the parent metal, resulting in a weld that is high in tensile strength and low in ductility. Depending on the joint design, weld procedure, amount of preheat, and final

heat treatment of the completed weld, the welded joint should have sufficient strength. However, the low ductility of the weld may cause failure of the joint during a stress application.

The iron-nickel-manganese electrode (ENiFe-C1-A) consists of 45% iron, 50% nickel, 1% carbon, and approximately 4% manganese. It is specially produced for welding cast irons. Tensile strength is not a problem with this electrode. The ductility of the weld and heat-affected zones is improved by the nickel content and the addition of carbon to the weld zone.

Material Preparation

The weld area should be cleaned and any dirt, rust, scale, or grease removed before the welding operation. Such contaminates will affect the deposition of the weld metal and may cause defects in the weld or weld interface. Castings with surface scale should be ground and the scale removed. Follow safety instructions when grinding. Always wear eye protection. Thermal-cut materials have oxidized metal on the face of the cut, providing a place for weld defects to form. These areas should be ground to bare metal before welding.

Grease burns and leaves a residue that causes porosity in the weld. Use degreasers to remove grease from the weld area, and follow recommended safety rules.

Cracks in castings are caused by stresses beyond the strength of the material. The end of the crack may not be visible but may be located with a penetrant test. Be sure dye and developer are completely removed from the crack before welding. Often, holes are drilled at each end

Steel Electrodes	Characteristics
EXXT-4	• Direct current, electrode positive (DCEP). • Self-shielding; no shielding gas required. • Globular arc mode of transfer. • Weld is hard and cannot be machined. Has little resistance to cracking.
EXXT-6	• DCEP. • CO_2 or 75% Ar/25% CO_2 shielding gas. • Globular arc mode of transfer. • Low hydrogen-type flux; therefore, weld is more crack-resistant than with T-4 electrode.
EXXT-7 EXXT-8	• Direct current, electrode negative (DCEN). • No shielding gas required. • In small diameters, can be used for all positions. • Welds are extremely hard, have very little crack-resistance, and are almost impossible to machine.
ENiFe-C1-A	• DCEP. • Shielding gas varies depending on electrode supplier. Manufacturer may specify use CO_2, use 98% Ar/2% O, or no shielding gas required.

of the crack to prevent further cracking during the heating, welding, and cooling operations.

Welding Procedures

Cast iron is melted, poured into a mold, and slowly cooled. Slow cooling allows stresses to relieve themselves and prevents cracks from forming. Regardless of the technique used, welding stresses the casting; therefore, the procedure must almost duplicate the original casting procedure, in terms of solidification of the metal and cooling of the casting.

Basic welding operations include:
- Repair of surface defects in the shell of the casting.
- Adding sections to complete an assembly design.
- Repair of a worn area or break in the casting part.

Preheat, Interpass, and Postheat

Preheating, interpass heating, and postheating may or may not be critical in welding a casting. Thin parts or attachment components made of other types of materials may not need preheating. However, a large casting with a thin part attached may require heating the large part to prevent cracking. A shallow skin defect may not need preheating; however, if this defect is located in a complex design area, preheat may be required.

There are no rules for heating requirements. For the most part, studying the weld to be made and using common sense will best determine whether to apply heat and for how long. Remember that material expands when heated, and it contracts (shrinks) when cooled. If the material cannot contract on cooling, stresses will form and cracks will result.

Preheating heats a local area or an entire part. For small local areas on the casting skin, preheat is not required unless moisture is present. Moisture can also be removed with alcohol, acetone, or air. If the weld is in a complex stressed area, full preheat is probably required. Full preheat ranges from 600°F (316°C) to 1000°F (538°C) for high-carbon material.

During welding, the interpass temperature of the entire part must not fall below the preheat temperature. If the part is postheated or stress-relieved after welding is completed, the casting may crack when the part is placed in the furnace.

Postheat is just as important as preheat. Postheating must be completed before the part cools below the interpass temperature. Heating blankets should be applied to the part to allow slow cooling, thereby eliminating stresses and potential cracking.

Whenever any type of heating is used, the part must be protected to prevent rapid cooling. Windy areas can cause cooling and cracking in the weld or the weld heat-affected zone.

Joint Preparation

When considering joint designs:
- Design the joint to compress during welding, **Figure 11-1.**
- Be aware that single V-groove welds shrink more than double V-groove welds, **Figure 11-2.**
- Design weld joints of equal thickness, **Figure 11-3.**
- Use fillet welds for minimum shrinkage.
- Butter component parts or joint sidewalls to reduce the amount of carbon pickup from the groove wall and to reduce the hardness of the weld. See **Figure 11-4.**

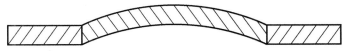

Figure 11-1. A domed piece of material will shrink during welding, reducing the possibility of the base material cracking.

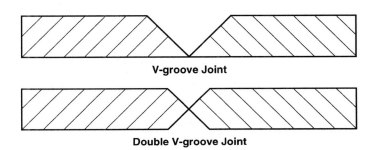

V-groove Joint

Double V-groove Joint

Figure 11-2. The V-groove joint requires twice as much filler material as the double V-groove joint.

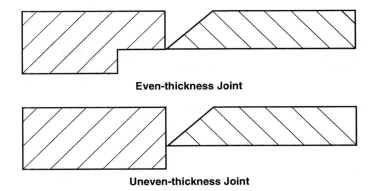

Even-thickness Joint

Uneven-thickness Joint

Figure 11-3. Even-thickness joints allow the material to shrink evenly. Uneven thickness joints will not cool at the same rate, causing stresses to build up in the joint and possible failure.

Figure 11-4. Buttering welds reduces the possibility of shrinkage cracks since the filler material has more elongation than the cast iron.

- Replace broken pieces with steel materials, if possible. Allow for weld shrinkage. See **Figure 11-5.**

Test Welds

The welding operation is dependent on a correct procedure for the deposition of the filler material. Typical welding parameters for various diameters of electrodes are listed in **Figure 11-6.**

Set your machine to produce sufficient amperage and voltage consistent with good fusion. Establish the machine setting by welding on a piece of scrap metal and

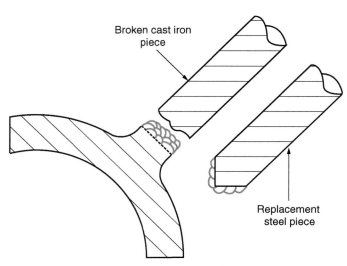

Figure 11-5. Grind the break area clean, inspect for cracks, and butter the end of the break with several layers of nickel filler material. The replacement piece can also be built up with nickel, if desired. Complete the weld with nickel filler material.

Welding Parameters			
Alloy Rods Corp. NICORE 55 electrode. Direct current, reverse polarity (DCRP); 98% Ar/2% O shielding gas at 20–45 cfh.			
Diameter	**Amps**	**Volts**	**Stickout**
0.035″	150–180	26–28	1/2″
0.045″	220–250	27–29	1/2″
1/16″	280–320	28–30	5/8″
3/32″	380–420	30–32	5/8″
Huntington Alloys NI-ROD FC 55 electrode. DCRP; CO_2 shielding gas at 30 cfh.			
Diameter	**Amps**	**Volts**	**Stickout**
0.078″	310	29	3/4″
0.093″	350	29	3/4″

Figure 11-6. These welding parameters are general; consult with the filler material manufacturer for specific information.

observing metal deposition. If you intend to preheat the actual weld, preheat the test part as well. The welding parameters will be lower for a preheated part than for a cold part. Once you have established machine parameters and variables, record the settings for future use. These settings will serve as a base procedure from which changes can be made.

Problem Areas

Several suggestions can help eliminate problems associated with welding cast irons using the FCAW process:

- Remove porosity in the weld by grinding or chipping, *not* by welding over it. When a weld is done, peen it immediately with a rounded tool, using rapid, moderate blows. This will expand the weld metal and reduce the amount of weld shrinkage.
- To prevent slag from moving ahead of the weld, slightly tilt the gun angle backhand. Always remove slag and wire-brush a completed weld before depositing another layer of material.
- Stringer beads work best by applying heat to a small area. Wide wash beads create more heat in the part with the greatest weld shrinkage and distortion, leading to heat-affected zone cracking.
- Backstep welding and skip welding help reduce heat input and cracking at the weld interface and heat-affected zones, **Figure 11-7.** Make short welds so you can peen the completed weld before it cools.
- When using the FCAW process, welders tend to make longer welds than when using other processes. This can be overcome by following the suggestions above. *Do not try to hurry the welding. Stay with your procedure.*
- Inspect each weld bead for cracks, lack of fusion, and porosity. Adjust the parameters to eliminate

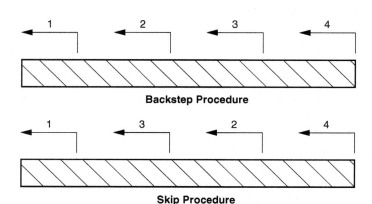

Figure 11-7. Backstep and skip welding reduce expansion and subsequent joint shrinkage during welding. These techniques are often used when the part cannot be heated to a higher temperature.

these conditions. The weld bead should not have a high crown. A high crown on the weld can cause cracking at the weld interface when stressed. Remove these areas and any undercut before allowing the casting to cool. If air-powered tools are used, do not allow the exhaust to flow onto the casting. Rapid chilling in a local area may cause cracking.

- If the procedure requires postheating, do so immediately. If a furnace operation is required for stress-relief of the weld, make sure the part is at an even temperature *before* it enters the furnace.

Facts about FCAW for Cast Irons

Study these general facts concerning the use of FCAW with cast irons:

- High voltage causes air contamination of the arc column and potential porosity of the weld.
- High voltage causes spatter.
- High current increases penetration and increases the mixture of the filler material and base material.
- Decreasing current reduces undercut.
- Decreasing stickout increases penetration.

- Increasing stickout can overheat the electrode and cause spatter.
- Stringer beads reduce shrinkage.
- Wash beads reduce porosity of the weld.
- Wash beads, since they are wider than stringer beads, require longer time periods for a particular length of weld. This increases the amount of shrinkage as the weld cools, producing shrinkage cracking.

Figure 11-8 and **Figure 11-9** show two applications of a flux cored welding electrode for the repair of large cast iron castings.

Review Questions

Please do not write in this text. Write your answers on a separate sheet of paper.

1. Cast iron has very high _____ and very low _____.
2. Small flakes of _____ form during the solidification of the gray iron in the cooling mold.
3. During the manufacture of malleable iron, the graphite changes into a _____ shape. Some alloys of this type of material are weldable.
4. The cast iron with the greatest ductility is called _____ cast iron.

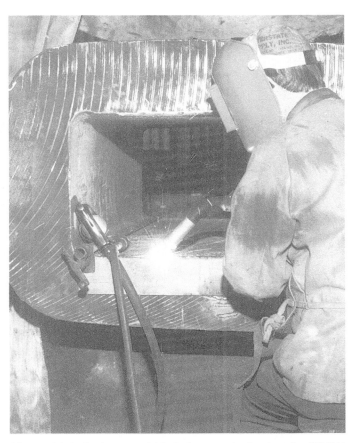

Figure 11-8. An ingot mold is being repaired using the FCAW process. (Inco Alloys International, Inc.)

Figure 11-9. This crane hoist has been rebuilt using the FCAW process. (Inco Alloys International, Inc.)

5. Excess carbon in spheroidal shapes rather than flakes gives the material higher _____ properties and good ductility.

6. Some steel electrodes are used for welding the cast irons. However, the best electrodes are _____ composition.

7. Materials that have been thermal-cut should always be ground to _____ prior to welding.

8. When cracks are found in the repair area, how can further cracking be prevented?

9. Compare shrinkage among single V-groove welds, double V-groove welds, and fillet welds.

10. If you preheat, you need _____ temperature control.

11. Buttering a component prior to making a groove weld usually results in a better weld with less _____.

12. Why should a finished weld be peened immediately before it cools?

Surfacing

Objectives

After studying this chapter, you will be able to:
- Describe the purpose of surfacing.
- Select the proper electrode based on usability and performance capabilities.
- Indicate how to safely prepare materials for welding.
- Describe the procedure for making surfacing welds.
- Consider temperature, joint preparation, filler material selection, equipment coordination, proper shielding gases, and test welds when establishing a welding procedure.
- Explain the criteria for determining whether surfacing is cost-effective.

Important Terms

buildup
buttering
cladding
hardfacing
stepover distance

Surfacing

Various metals are surfaced or covered with other metals to protect them from another material that could cause wear or deterioration of the original part. Wear includes abrasion, impact, heat, corrosion, or a combination of these factors. The part may be made undersize, surfaced, then machined to size as a new part, or it may be rebuilt from a used part. In any case, the surfacing operation can extend the life of the part at considerable cost savings.

Hardfacing is a special form of surfacing applied to make a part harder as a means of reducing wear. Not all surfacing welds are applied to make the part harder. The welds may be applied for any of the factors stated above. See **Figure 12-1.**

Several other terms are used to describe the surfacing operation. *Buttering* a part is the process of surfacing an area where materials are to be joined and a dissimilar material is to be used for the joining. This type of weld may also be used when two component parts are to be welded and heat-treated to move the joint stress from the components. Buttering is also used to obtain specific chemical values in the final weld. Such is the case when applying stainless steel onto carbon steel, and the final deposit needs a considerably lower carbon content. In this case, a different stainless steel is used for the buttering operation than is used for the final joining weld.

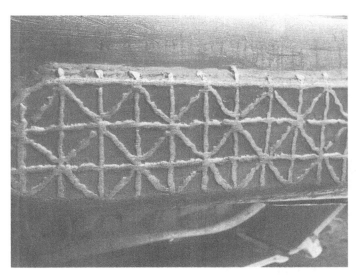

Figure 12-1. This tractor blade arm was hardfaced for protection during loading.

Cladding is the application of a material that provides a corrosion-resistant surface. *Buildup* welds are made to restore worn parts to original dimensions or to apply the surfacing material.

Base Materials

Cast irons, low- and high-alloy steels, stainless steels, and manganese steels are the most common materials surfaced with the FCAW process.

The mining, steelmaking, railroad, and construction industries use many types of base materials for the manufacture of their products and equipment. Many have been specially developed or modified for use in a specific industry application.

Filler Materials

Some filler materials fit into an AWS specification for a particular material. However, many of the surfacing materials have been developed by electrode manufacturers for specific applications and are not classified by AWS. These proprietary electrodes are listed by a number or a trade name.

An electrode may be designed for buildup, impact resistance, hardfacing, or corrosion- and heat-protection. The various electrode and chemical elements used in their manufacture are described next.

Buildup Electrodes

Buildup electrodes may or may not match the parent metal chemical analysis and mechanical values developed as a base for the final surfacing material.

Carbon-steel electrodes consists of the following elements:

- Iron-base core.
- Carbon.
- Manganese.
- Silicon.
- Chromium.
- Molybdenum.

Total alloy content is low and has good impact strength and machinability for application of the final surfacing material. Carbon-steel electrodes are used for buildup that may require multiple layers. Buildup layers should be made within one or two layers of the final dimension. The actual surfacing material is then applied for the final surfacing operation. These electrodes are designed for welds on all weldable carbon and low-alloy steels. The welds have a high resistance to deformation.

Manganese steel electrodes consists of:

- Iron base core.
- Carbon.
- Manganese.
- Nickel.

Total alloy content is high due to the increased strength level of the base material. The electrode may be used to join manganese steels, repair worn areas, and build up for surfacing. The deposit work-hardens to a high tensile strength and has good impact properties.

Buildup electrodes for welding on cast iron are limited to the steel electrodes in some areas and the iron-nickel electrodes listed in Chapter 10. The use of steel for a buildup electrode creates extreme hardness in the weld area due to carbon pickup from the cast iron. The material then cracks along or across the weld, leaving a poor base for applying the surfacing material. Iron-nickel-manganese steel electrodes are strongly recommended for this application.

Surfacing Electrodes

Surfacing materials can usually be applied to any of the buildup materials and many base materials. The main objective is to retain the values of the surfacing material after the welding operation. This is accomplished by using the correct procedures to minimize the depth of penetration on all buildup and surfacing welds, and by properly adding the final surfacing material. Incorrect procedures or excessive dilution will nullify the weld quality and lower the surfacing protection desired. Contact the surfacing material manufacturer for the proper welding procedure.

During surfacing operations on construction equipment using extremely high-alloy-content electrodes, the weld may have a tendency to crack either across or in line with the weld. This is caused by rapid cooling of the weld and extreme hardness of the deposited metal. In most cases, these cracks do not have an adverse effect on the usability of the completed weld.

Surfacing electrode composition

Through experience and physical testing, each manufacturer has established the chemical values of surfacing electrodes and the procedures for applying them. The electrodes and chemical requirements in **Figure 12-2** are general and serve only as a guide for selection of the desired electrode. The manufacturer has thoroughly tested the welding material and found the recommended electrode works very satisfactorily for the intended application.

Material Preparation

The weld area should be cleaned and any dirt, rust, scale, or grease removed before welding. Such contaminates will affect the deposition of the weld metal and may cause defects in the weld or weld interface. Castings with surface scale should be ground and the scale removed. Thermal-cut materials have oxidized metal on the face of

Surfacing Electrode Composition	
Type of Electrode	**Application**
Stainless steel	Used as an overlay for the carbon steels for corrosion protection. Type 309L S/S is used for the first layer, and type 316L is used for the final surfacing weld.
High percentages of manganese and chromium	Used on materials requiring protection from moderate to severe abrasion. These electrodes generally work-harden in use, and cannot be flame cut or machined after welding.
Nickel base with manganese, chromium, tungsten, and molybdenum	Used for resistance to high heat and corrosion, with good impact and abrasion values.
Iron base with small amounts of tungsten carbides within the core	Have excellent resistance to earth wear.

Figure 12-2. Chemical composition of surfacing electrodes and their applications.

the cut, providing a place for weld defects to form. These areas should be ground to bare metal before welding.

Grease will burn and leave a residue that causes porosity in the weld. Use degreasers to remove grease from the weld area, and follow recommended safety rules. In areas where parts of previous weld deposits remain, the old material should be removed to bare material. Inspect the material thoroughly, and repair cracks before applying buildup or surfacing materials. Follow safety instructions when grinding or using degreasers. Always wear eye protection.

Welding Procedures

Each type of surfacing requires a different procedure, depending on the base material and type of welding electrode used. **Figure 12-3** shows a button-type pattern in an industrial application.

The object of surfacing is to deposit the material into the base or buildup material with a minimum amount of dilution to maintain the proper metallurgical content of the filler material. The manufacturer of the filler material specifies the type of current and shielding

Figure 12-3. The welder is making button-type welds to reduce wear on the blade. (Eutectic Corp.)

gas (if required). The amperage should be sufficient to maintain a steady arc and good pool control. The electrode should be almost perpendicular to the weld, with only a slight amount of drag angle. This will reduce penetration; however, slag may flow forward and become entrapped if the arc is too long. High-arc voltages and long stickouts are fine for reducing penetration and minimizing dilution, but they tend to cause lack of fusion and slag entrapment.

The amount of dilution into the base material has a direct effect on chemical content of the final pass. To establish the amount of dilution of the weld metal, a test can be made to determine the actual welding parameters and the final results of the weld. Many surfacing electrodes limit the final weld to a single layer to maintain the desired surfacing qualities. During the test period, all welding parameters must be recorded so the procedure can be duplicated for the actual weld. If the desired qualities are not achieved during the test period, another test must be made with another set of welding parameters.

Surfacing welds are best applied in the flat position to limit heat input and impart the best bead shape to the completed weld. After the first pass is made, the remaining passes are usually located on half of the completed weld and half of the parent metal. The distance moved over to make this next weld is called the ***stepover distance.*** It must be maintained until all the required welds are completed. **Figure 12-4** shows a badly worn track roller to be rebuilt by welding. **Figure 12-5** shows the completed weld on the worn track roller. Note how the welds are moved over on each circular pass.

Oscillated welds are made by moving the gun across the weld path in a set pattern. These types of welds are made more slowly. More distortion occurs, and base material heat can affect the amount of dilution. Oscillated welding is easily controlled on an automatic machine where the oscillation parameters can be determined and controlled. Welders using semiautomatic equipment should make test welds with the proposed procedure and production equipment to verify the actual welding parameters to be used.

Each weld of the overlay should be made to a specific procedure to maintain the chemical limits of the final weld. A weld that does not meet this requirement will fail in service, resulting in excessive repair costs or jeopardizing the safety of the equipment operator.

Figure 12-4. This tractor roller has been worn to the maximum allowance. All welding surfaces have been cleaned, and the unit is ready for welding.

Figure 12-5. A dual welding head is welding this roller. To reduce smoke from the operation, a neutral flux is used with a FCAW filler electrode. With automatic stepover programmed into the operation, welding can continue until the layer is complete.

Preheat, Interpass, and Postheat

The temperatures for various types of surfacing operations are dependent on many factors, including:

- Type of base material.
- Thickness of base material.
- Complexity of weldment design and type of weld joint.
- Type of filler materials to be used.
- Mechanical and chemical values of the completed weld.
- Final heat treatment of the surfaced part.

Other areas must be considered before welding the part. Some metals cannot be heated beyond a certain temperature—the heat will destroy the grain structure of the base material and cause the weldment to fail. Other metals must be postheated to prevent hardening of the final welds and cracking as the weld metal cools.

The following conditions for heating operations generally apply:

- Austenitic manganese steels are *not* heated.
- Chrome-nickel stainless steels are *not* heated.
- Low-carbon steels are *not* heated.
- Cast irons are heated.
- Extremely cold materials are heated. (Thermal shock may cause cracking of the base material or the weld.)
- High-chromium steels are heated.
- Medium- and high-carbon steels are heated.
- Very hard surfacing deposits are heated.

In general, if the part is to be heated, then the entire heat-affected zone requires heat. Complex design parts must be heated throughout to prevent cooling stresses from forming. If preheating is required, interpass heating and postheating are done also. Interpass temperature should be maintained until the welding operation is complete. The final heating should be started before the

Figure 12-6. A preheat calculator can be used to determine the heating temperatures required for surfacing carbon and low-alloy steels. (Lincoln Electric Co.)

part drops below the interpass temperature. Final cooling should be done by covering the part with an insulating material or a blanket, or placing it in a furnace to prevent rapid cooling. Rapid cooling results in internal or external stresses that may cause cracking of the weld or the weldment.

Several companies have developed preheat and interpass temperature calculators for determining required temperatures when welding and surfacing carbon and low-alloy steels. See **Figure 12-6.**

Welding Techniques

Many completed hardfacing welds on construction equipment have patterns with no definite designs that cover only part of the wearing surface. See **Figure 12-7.**

Figure 12-7. Many patterns were used to hardface this dragline bucket. (Lincoln Electric Co.)

Labor and materials would be too costly to cover the entire surface area. The weld patterns act as dams, while the "open" areas between the weld passes fill with dirt or sand, making welds unnecessary. Another consideration is hard material that does not compact into the nonwelded areas. All types of welding designs are used to protect the base material from wear depending on the type of earth being moved. Design may include buttons, circles, curved lines, zigzagged lines, and straight lines with large and small gaps.

To ensure the welds will effectively prevent wearing of the part, take into account the final use of the equipment when choosing a pattern. Applications of hardfacing on construction and industrial equipment are shown in **Figures 12-8, 12-9,** and **12-10.** Note the patterns used and the amount of the original area left uncovered.

Type of surfacing material, weld deposition time, and final use of the equipment are three factors that determine if a welding operation is cost-effective. If the cost of surfacing is too high, a new part may be used and the old, worn-out part thrown away. This is especially true in rework on tractors used for earthmoving. The front teeth on the bottom of the blade typically are not surfaced because the cost is too high. Instead, formed plates covering the entire part are applied. Welding is used only to attach the new plate to the old one.

Electrode and labor costs are high. However, filler materials for FCAW are considerably less expensive than filler materials for SMAW. When equipment is repaired in the field, the use of portable equipment reduces downtime and lowers costs for the operator.

Figure 12-11 shows a roll prepared for weld surfacing. The weld is to be applied by an automatic arc

Figure 12-9. The bottom edge of this blade is hardfaced to protect it from wear.

Figure 12-10. The weld pattern traps dirt or sand in the pockets. The areas are protected as if they were welded.

welding process using a FCAW filler material and submerged arc welding neutral flux. The equipment and tooling is adaptable for the application of this surfacing material. See **Figure 12-12** for the welding operation.

Application of the material by a two-head machine is easy and practical. The results have exceptional quality with a very even crown from end to end. Smoke is significantly reduced by using a submerged arc welding flux in addition to the FCAW filler material.

Sometimes a surfacing operation is not practical because of the design of the part. In these cases, wear plates are used and attached by welding, **Figure 12-13.**

Figure 12-8. Bucket housings for these front blade teeth were hardfaced. Note the entire front blade is new; it is a replacement part and has not been hardfaced.

Figure 12-11. The roller shown here has been prepared for welding and is mounted on a special machine for the surfacing operation. (Lincoln Electric Co.)

Figure 12-12. Welding done with two heads reduces time. Weld quality is excellent when welding is done under controlled conditions.

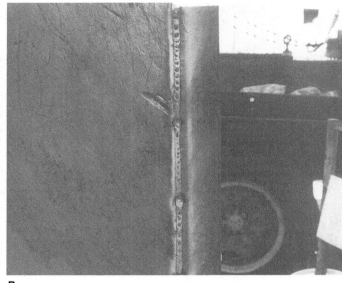

A

B

Figure 12-13. Typical wear plate installations for protection of tractor blades. A—Wear plates are used when sufficient metal is not available as a base for hardfacing. B—The wear plate was welded uphill in the vertical position. The weld was made with a backstep technique to reduce heat input and warpage.

Surfacing Area Safety

Safety must be practiced in the surfacing area to prevent accidents and injuries. Working around heavy machinery requires being alert to others working near you. Once your hood is down, you cannot see your surroundings. Let others know you are there so they can help you, if needed. If possible, have a person work with you at all times. Make sure you have all the equipment necessary to complete the task. Keep a fire extinguisher handy when working near a gasoline engine, a diesel engine, or a fuel tank.

Test Welds

The actual welding operation should match the test procedure as closely as possible. The test procedure determines the correct welding parameters and variables necessary to complete the task and obtain a satisfactory weld. The welder who develops the procedure should carry out the actual welding or should instruct the assigned welder in proper operation of the procedure. A test weld should be made after the machine is set up to confirm the settings are correct. A change in electrode diameter, welding voltage, stickout, or travel speed may affect the deposition characteristics and cause the weld to fail.

Conclusions

Surfacing offers many industries the opportunity to reduce maintenance costs on a variety of equipment. Replacement costs have increased so much that welding is now being used to surface new parts before they are placed in service to increase their useful life. As you work in the welding field, be attentive to new areas of surfacing in which you could become a welding specialist, thereby increasing your employment potential.

Review Questions

Please do not write in this text. Write your answers on a separate sheet of paper.
1. What is the purpose of surfacing?
2. What are the four causes of wear?
3. When welding two different material components together, a(n) _____ material is used to make the components compatible.
4. The object of surfacing is to deposit the material into the base or buildup material with a minimum amount of _____ to maintain the _____.
5. What type of electrode is used for the filler material when welding a component that has considerable wear?
6. What two types of buildup electrodes are used for cast iron?
7. Holding a long electrode extension will _____ penetration into the parent metal.
8. With surfacing, the welding electrode should be held _____ to the workpiece, and all welding should be done in the _____ position.
9. Oscillated welds are made slowly. As a result, more _____ occurs and the amount of _____ is affected.
10. Weld patterns are often designed to make _____ to hold dirt, sand, or other soft materials.
11. What three factors determine if a welding operation is cost-effective?

CHAPTER 13 Procedure and Welder Qualification Test

Objectives

After studying this chapter, you will be able to:
- Identify the AWS specification for qualification of welding procedures and welders.
- Summarize the basic requirements for welder certification.
- Explain how specifications govern the quality of the completed weldment.

Important Terms

macroetch test
nonprequalified joints
prequalified joints
Procedure Qualification Record (PQR)
side bend test
welder qualification
Welding Procedure Specification (WPS)

Author's Note to the Reader

This chapter outlines the AWS specification for qualification of welding procedures and welders, giving the basic requirements for welder certification. There are many other welding tests within the code, as well as other codes and specifications that govern the welding of various types of joints and metals. A study of the various welding specifications will provide the correct information for welding and inspecting the various joints properly. Keep in mind that specifications can change often as new materials, processes, variations, and inspection procedures are introduced.

Specifications govern the quality of the completed weldment. You must carefully consider each weld you make or for which you are responsible. The safety of many people is at stake if the weld is not made to the rules of the specification. Shortcuts in quality cannot be tolerated in order to lower production costs or decrease production time. Welders must know the requirements of the specification and make each weld to those requirements.

The American Welding Society authorizes testing laboratories to conduct the *D1.1 Structural Welding Code—Steel* test, and to qualify welders who demonstrate proficiency in a specific welding process. For a list of authorized laboratories offering the qualification test, contact the AWS at:

American Welding Society
550 N.W. LeJeune Rd.
Miami, FL 33126

Specifications for the welding of numerous products and materials are available for purchase through AWS. Membership in the organization is also available.

Buildings, structures, ships, pressure vessels, nuclear reactors, and many other applications require sound welds made to a proven procedure by a qualified welder. Several specifications are currently in use for these application areas and are available for purchase. Two major codes governing welding are the *American Society of Mechanical Engineers (ASME) Boiler Code* and the *American Welding Society (AWS) Structural Welding Code.*

American Society of Mechanical Engineers Boiler Code

Section 9 of the Boiler Code governs the welding of pressure vessels, boilers, and atomic reactors, and requires testing and qualification of all welding procedures and welders. All types of base materials, filler materials, and welding processes are permitted under the code. Weld tests are required to verify the strength and quality of the weld. The weld test parameters and the variables to be used are entered on a *Welding Procedure*

Specification (WPS) form. Nondestructive and destructive tests are performed, and the results recorded on a *Procedure Qualification Record (PQR).* If the results are acceptable, the procedure is considered "qualified" until weld quality is unacceptable, joint design is changed, or parameters or variables exceed the specification limits. Welder qualification is obtained using an approved WPS.

American Welding Society Structural Welding Code

AWS D1.1 Structural Welding Code—Steel is used for the construction of bridges, buildings, and similar structures using only carbon and low-alloy steels, with some limitations on welding processes. The code permits use of specified weld joint designs with various processes without requiring qualification tests for mechanical values when base material and filler material combinations are specified in the code.

These joint types are called *prequalified joints.* **Figure 13-1** shows a typical joint with the setup requirements for FCAW and allowable welding positions. The manufacturer completes a Welding Procedure Specification form, indicating the welding parameters and variables to make the weld. To ensure the quality of the weld matches the quality of the proposed procedure, a welding test is completed using the WPS. The completed test is subjected to nondestructive and destructive testing. The results are recorded on the PQR. The procedure is considered qualified if the test results are satisfactory.

When welding *nonprequalified joints,* or materials that have not been approved, a WPS qualification test is required, and the results are indicated on a PQR. Results include a mechanical test for metal strength, a bend test to prove the welding procedure, and other tests as required.

Procedure qualifications are required for all joint types and positions for both prequalified and nonprequalified joints. Changes beyond the code require complete retesting of the procedure and a new WPS and PQR. Production welding is allowed using only a proven WPS and the variables specified in the code.

AWS Welder Performance Qualification Test

AWS requires qualification for welders, welding operators, and tackers welding under code requirements. Qualification tests for FCAW welders include:
- Plate groove welds and position of the weld.
- Pipe groove welds and position of the weld.

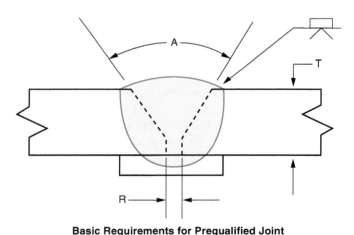

Basic Requirements for Prequalified Joint

Process Types and Limitations
T = Material Thickness Limitations

Groove Preparation
A = Bevel angle tolerances
R = Root opening tolerances

Gas Shielding Requirements

Permitted Welding Positions
- Flat
- Horizontal
- Vertical
- Overhead

Figure 13-1. AWS prequalified weld joint for V-groove welds.

Welder Qualification—Type and Position Limitations					
Qualification Test		Type of Weld and Position of Welding Qualified*			
		Plate		Pipe	
Weld	Plate or pipe positions	Groove	Fillet	Groove	Fillet
Plate-groove	1G	F	F,H	F	F,H
	2G	F,H	F,H	F,H	F,H
	3G	F,H,V	F,H,V	F,H,V	F,H
	4G	F,OH	F,H,OH		F
	3G & 4G	All	All		F,H
Plate-fillet	1F		F		F
	2F		F,H		F,H
	3F		F,H,V		
	4F		F,H,OH		
	3F & 4F		All		
Pipe-groove	1G	F	F,H	F	F,H
	2G	F,H	F,H	F,H	F,H
	5G	F,V,OH	F,V,OH	F,V,OH	F,V,OH
	6G				
	2G & 5G				
	6GR	All	All	All	All
	6GR		All		All
Pipe-fillet	1F		F		F
	2F		F,H		F,H
	2F Rolled		F,H		F,H
	4F		F,H,OH		F,H,OH
	4F & 5F		All		All
*Positions of welding: F=flat, H=horizontal, V=vertical, OH=overhead					

Figure 13-2. AWS welder qualification limitations.

- Plate fillet welds and position of the weld.
- Plate plug and slot welds and position of the weld.
- Pipe fillet welds and position of the weld.

Some tests also require qualification for the minimum and maximum thickness of the material to be welded.

Welders who qualify for the plate test are also qualified for fillet welds with the same thickness and position restriction. **Figure 13-2** shows the type and position limitations for plate groove and plate fillet welds.

Welder Qualification Weld Test Types and Positions

Groove weld test plates for unlimited welding thickness are 1″ thick and qualify for welding 3/16″ thickness and greater. Groove-weld test plates for limited thickness are 3/8″ thick and qualify for welding 3/16″ minimum and 3/4″ maximum.

Fillet weld test plates are 1/2″ thick and qualify all thicknesses over 3/16″. To weld thicknesses below 3/16″ on plate and fillet welds, use AWS specification *D1.3 Structural Welding Code—Sheet Steel.* See **Figures 13-3** through **13-7** for test positions, and specimen requirements and thicknesses.

Weld Test Plate Preparation and Dimensions

The test plate for welder qualification for unlimited thickness groove welds is shown in **Figure 13-8.** An optional test plate for unlimited thickness horizontal position welder qualification groove weld test is shown in **Figure 13-9.**

The joint design for tubular butt joints for welder qualification is shown in **Figure 13-10.** The fillet weld plate test for welder qualification is shown in **Figure 13-11.**

Flat

Vertical

Horizontal

Overhead

Figure 13-3. Test positions for plate groove welds.

Flat 1G (Pipe rotated during welding)

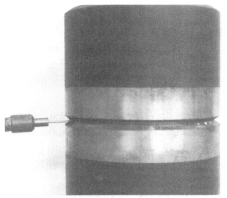

Horizontal 2G (Pipe not rotated during welding)

Vertical 5G (Pipe not rotated during welding)

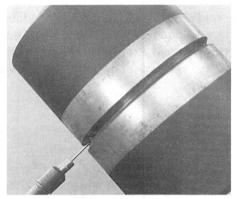

6G (45°) (Pipe not rotated during welding)

Figure 13-4. Test positions for pipe or tubing groove welds.

Flat 1F

Horizontal 2F

Overhead 4F

Vertical 3F

Figure 13-5. Test positions for plate fillet welds.

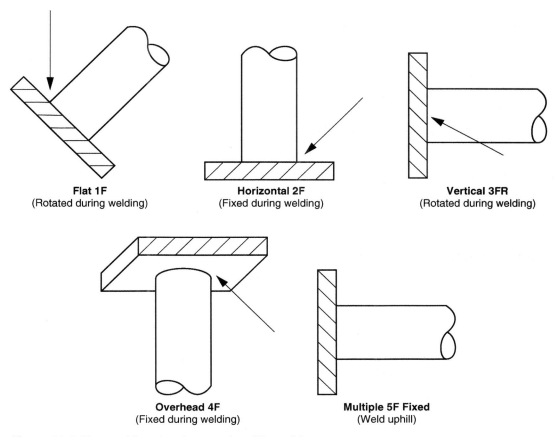

Figure 13-6. Test positions for pipes to plate fillet welds.

Number and Type of Specimens and Range of Thickness Qualified-Welder Qualification											
Plate Test			**Number of Specimens**						**Plate Thickness Qualified**		**Dihedral Thickness Qualified**
Type of weld	Thickness of test plate (T) as welded	Visual inspection	Bend Test		Side	T-joint break	Macroetch test				
			Face	Root				Min.	Max.	Min.	Max.
Groove	3/8″	Yes	1	1	—	—	—	1/8″	3/4″		
Groove	3/8<T<1	Yes	—	—	2	—	—	1/8″	2T		
Groove	1″ or over	Yes	—	—	2	—	—	1/8″	Unlimited		
Fillet Option No. 1	1/2″	Yes	—	—	—	1	1	1/8″	Unlimited	60°	135°
Fillet Option No. 2	3/8″	Yes	—	2	—	—	—	1/8″	Unlimited	60°	135°
Plug	3/8″	Yes	—	—	—	—	2	1/8″	Unlimited		

Figure 13-7. Welder qualification test specimen requirements and thickness qualified.

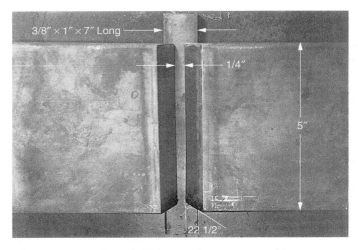

Figure 13-8. Unlimited thickness plate groove weld test.

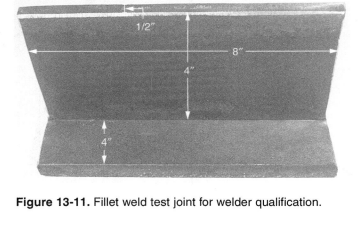

Figure 13-11. Fillet weld test joint for welder qualification.

Figure 13-9. Optional unlimited thickness horizontal weld position for plate groove weld test. This groove design can be used in place of the V-groove design.

Qualification Test

The fabricators of many welded products utilize AWS D1.1 to control the qualification of welders and ensure the quality of welding on their products is satisfactory. The procedures described next can be used for welding procedure testing and welder qualification.

Welder qualification is based on the ability of the welder to utilize a proven welding procedure using a specified process and the proper techniques to complete a satisfactory weld. The welder is required to utilize all the data on the Welding Procedure Specification. Welder certification test data for welding parameters and variables is not to be used for production welding.

Welder Groove Weld Qualification Test

This test duplicates an actual AWS certification test for FCAW on steel plate for unlimited thickness qualification. Use the welder groove-weld test WPS shown in **Figure 13-12.** The material should be a mild steel, prepared as shown in Figure 13-8. If a radiographic test for quality is to be done rather than destructive testing, guided bend tests are not required, and the backing bar should be 3″ wide instead of the normal 1″ wide steel bar.

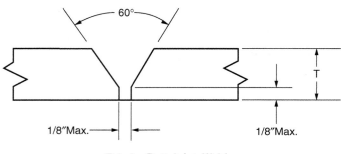

Tubular Butt Joint–Welder Qualification–without Backing

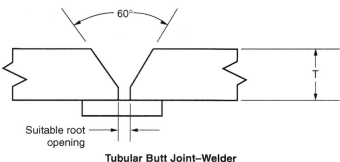

Tubular Butt Joint–Welder Qualification–with Backing

Figure 13-10. Joint designs for tubular butt joints for welder qualification.

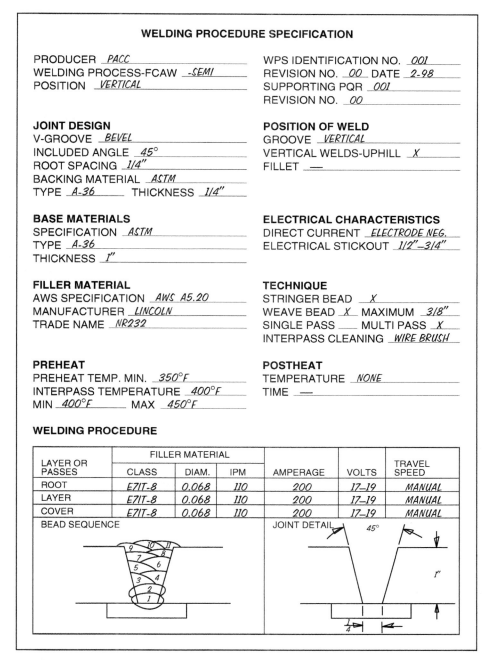

WELDING PROCEDURE SPECIFICATION

PRODUCER *PACC*

WELDING PROCESS-FCAW *-SEMI*

POSITION *VERTICAL*

WPS IDENTIFICATION NO. *001*

REVISION NO. *00* DATE *2-98*

SUPPORTING PQR *001*

REVISION NO. *00*

JOINT DESIGN

V-GROOVE *BEVEL*

INCLUDED ANGLE *45°*

ROOT SPACING *1/4"*

BACKING MATERIAL *ASTM*

TYPE *A-36* THICKNESS *1/4"*

POSITION OF WELD

GROOVE *VERTICAL*

VERTICAL WELDS-UPHILL *X*

FILLET *—*

BASE MATERIALS

SPECIFICATION *ASTM*

TYPE *A-36*

THICKNESS *1"*

ELECTRICAL CHARACTERISTICS

DIRECT CURRENT *ELECTRODE NEG.*

ELECTRICAL STICKOUT *1/2"–3/4"*

FILLER MATERIAL

AWS SPECIFICATION *AWS A5.20*

MANUFACTURER *LINCOLN*

TRADE NAME *NR232*

TECHNIQUE

STRINGER BEAD *X*

WEAVE BEAD *X* MAXIMUM *3/8"*

SINGLE PASS ___ MULTI PASS *X*

INTERPASS CLEANING *WIRE BRUSH*

PREHEAT

PREHEAT TEMP. MIN. *350°F*

INTERPASS TEMPERATURE *400°F*

MIN *400°F* MAX *450°F*

POSTHEAT

TEMPERATURE *NONE*

TIME *—*

WELDING PROCEDURE

LAYER OR PASSES	FILLER MATERIAL			AMPERAGE	VOLTS	TRAVEL SPEED
	CLASS	DIAM.	IPM			
ROOT	E71T-8	0.068	110	200	17–19	MANUAL
LAYER	E71T-8	0.068	110	200	17–19	MANUAL
COVER	E71T-8	0.068	110	200	17–19	MANUAL

BEAD SEQUENCE   JOINT DETAIL   45°   1"   1/4

Figure 13-12. Welding Procedure Specification for welder groove weld test.

Remove all slag and wire brush the material, **Figure 13-13.** Assemble the plates on the backing bar with a 1/4″ gap, and tackweld each side piece to the backing bar, **Figure 13-14.** Turn the assembly over, and make a complete weld on each side of the backing bar. This weld will reduce the amount of distortion of the completed assembly and must be done if the bend test specimens are to be oxyacetylene-cut after welding is completed. See **Figure 13-15.**

Place the completed assembly in the desired position for welding. (Refer to Figure 13-3 for groove weld test

Figure 13-13. Wire brush or grind the entire material welding area to remove scale or foreign materials before tackwelding.

Figure 13-14. Tackweld plates on each end of the backing bar with the proper gap dimension. Make sure the plates are level and in full contact with the backing bar.

Figure 13-15. Each side of the backing bar is welded to minimize distortion during groove welding.

positions.) Complete the weld per the WPS, using the proper welding parameters. Make sure the crown dimensions are within specifications. See **Figure 13-16** for a

Figure 13-16. When welding a groove weld, keep the layers level, and allow space for the final pass to tie into the preceding pass and plate sidewall. The cover pass should extend over the groove joint equally on each side. This sequence may be used in any position.

partial test made to the WPS showing bead placement. A completed weld test plate is shown in **Figure 13-17.**

During welding, watch for possible weld test failure. Problems can be prevented by following these suggestions:

- Make a test weld after you have set up the machine to be sure all the settings are correct.
- Use the proper lead or drag angle on the torch.
- Maintain the proper stickout throughout the weld.
- Maintain the welding parameters within the machine's duty cycle limits.
- Use stringer or wash beads as directed by the WPS.
- Remove slag from each pass; wire brush and inspect.
- Take time to do the job right. Know the final crown height dimensions, and stay within these limits.

Welding Inspection

After welding is completed, the test is evaluated to the specification for weld bead profile and visual defects. Study **Figure 13-18.** Cracks are not acceptable in any welded test plate and are cause for rejection. Undercut next to the outer edge of the weld bead is limited to 1/32″. Measuring such a small distance is difficult, and the eye can be deceived. Gauges are the best tool for measuring the depth of undercut, **Figure 13-19.**

Destructive Testing

Two 3/8″-wide coupons (specimens) are required from each butt weld test plate for the side bend test. If the coupons are cut by a bandsaw, they may be 3/8″. If they are cut with an oxyacetylene torch, the coupon must be 5/8″ wide, and 1/8″ must be machined from each side to achieve a 3/8″ coupon. See **Figure 13-20.**

Remove the backing strip with a gouging torch, and grind any remaining material from the root side. Two oxyacetylene-cut and machined coupons with the backing bar removed are shown in **Figure 13-21.** Polish

Figure 13-17. Completed groove weld test ready for nondestructive and destructive testing.

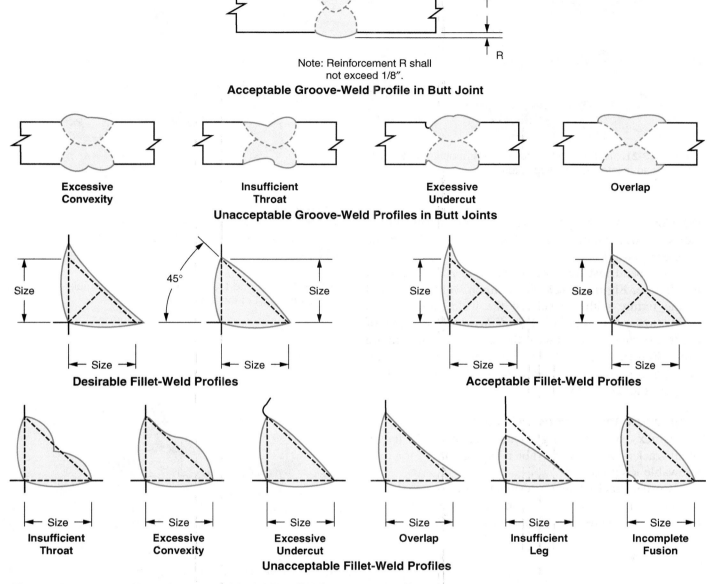

Note: Reinforcement R shall not exceed 1/8″.

Acceptable Groove-Weld Profile in Butt Joint

Excessive Convexity	Insufficient Throat	Excessive Undercut	Overlap

Unacceptable Groove-Weld Profiles in Butt Joints

Desirable Fillet-Weld Profiles

Acceptable Fillet-Weld Profiles

Insufficient Throat	Excessive Convexity	Excessive Undercut	Overlap	Insufficient Leg	Incomplete Fusion

Unacceptable Fillet-Weld Profiles

Figure 13-18. Acceptable and unacceptable weld profiles for groove and fillet welds.

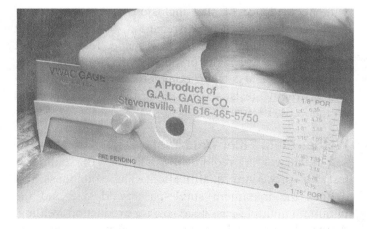

Figure 13-19. Inspection tool used to check undercut at the weld edge. The amount of undercut is shown on the scale. (G.A.L. Gage Co.)

Figure 13-20. V-groove weld test layout for cutting bend test coupons.

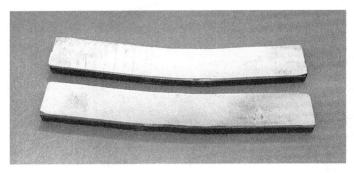

Figure 13-21. The coupons have been machined to 3/8" thickness to fit into the bending fixture.

Figure 13-23. Bend test coupons are bent completely around the plunger part of the test fixture.

the sides of the coupon, and round the edges. Grind excess crown, and remove any backing material from the base of the weld.

The bend test must be made in an AWS-approved test fixture, **Figure 13-22.** The coupon weld is centered in the fixture with the side of the weld cross section facing upward. This is called a *side bend test.* The ram of the test fixture is closed until the legs of the coupon are even, **Figure 13-23.** The coupons are evaluated after bending for cracks on the outer surface of the bent coupon, **Figure 13-24.**

Inspection and Rejection Criteria

Cracks over 1/8" long are unacceptable. The length of all cracks under 1/8" must be added together for a total allowable length of 3/8".

Corner cracks must be evaluated for visual slag or inclusions. Defects resulting from these areas are limited to 1/8". Specimens with corner cracks exceeding 1/4",

Figure 13-24. The outer surface of the test coupons are inspected for cracks.

with no evidence of slag inclusions or other types of fusion discontinuities, are disregarded, and a replacement test from the original weldment is made. Test plates that are to be radiographed for quality are evaluated using code requirements. Upon completion of nondestructive and destructive tests, acceptance or rejection is indicated on the PQR, **Figure 13-25.**

Welder Plate Fillet Weld Test

Fillet welding requires a Qualified WPS soundness test for procedure qualification. The soundness test defines the maximum single-pass weld size and the minimum multiple-pass bead size for production welding. A typical soundness test is shown in **Figure 13-26** using the WPS in **Figure 13-27** and the PQR test shown in **Figure 13-28.**

Figure 13-22. AWS-approved bend test fixture for butt weld coupons.

PROCEDURE QUALIFICATION RECORD

DATE _2-98_____ WPS NO. _001____ REV. NO. _00____ PQR NO. _001_____

GUIDED BEND TEST

Specimen Number	Type of Bend	Result	Remarks
1	SIDE	SATISF.	
2	SIDE	SATISF.	

PLATE WELD TEST   **FILLET WELD TEST**

APPEARANCE _SATISFACTORY_____   _____

UNDERCUT _SATISFACTORY_____   _____

CONVEXITY _SATISFACTORY_____   _____

CONCAVITY _SATISFACTORY_____   _____

WELD SIZE _SATISFACTORY_____   _____

MACRO ETCH _N. R._____   # 1 _____ # 2 _____ # 3 _____

BREAK TEST _N. R._____   _____

DATE TEST _2-98_____ ACCEPTANCE _O.K._____ REJECTION _____

PRODUCER _PCC_____

TITLE _____

DATE _2-98_____

Figure 13-25. After testing is done, the PQR is completed and indicates either acceptance or rejection.

A

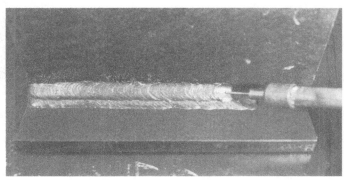

B

Figure 13-26. A fillet weld soundness test is made to determine weld procedures for maximum size single-pass and minimum size multiple-pass fillet welds. A—Single-pass weld. B—Multiple-pass weld.

WELDING PROCEDURE SPECIFICATION

PRODUCER *PCC*
WELDING PROCESS-FCAW *X*
POSITION *HORIZONTAL*

WPS IDENTIFICATION NO. *002*
REVISION NO. *00* DATE *2-98*
SUPPORTING PQR
REVISION NO.

JOINT DESIGN
V-GROOVE
INCLUDED ANGLE
ROOT SPACING
BACKING MATERIAL
TYPE _____ THICKNESS _____

POSITION OF WELD
GROOVE —
VERTICAL WELDS-UPHILL —
FILLET *HORIZONTAL*

BASE MATERIALS
SPECIFICATION *ASTM*
TYPE *A 36*
THICKNESS *1/2"*

ELECTRICAL CHARACTERISTICS
DIRECT CURRENT *ELECTRODE NEG.*
ELECTRICAL STICKOUT *3/4"*

FILLER MATERIAL
AWS SPECIFICATION *AWS*
MANUFACTURER *LINCOLN*
TRADE NAME *NR211 MP*

TECHNIQUE
STRINGER BEAD *ALL PASSES*
WEAVE BEAD — MAXIMUM —
SINGLE PASS *X* MULTI PASS *X*
INTERPASS CLEANING *WIRE BRUSH*

PREHEAT
PREHEAT TEMP. MIN. —
INTERPASS TEMPERATURE —
MIN —   MAX —

POSTHEAT
TEMPERATURE —
TIME —

WELDING PROCEDURE

LAYER OR PASSES	FILLER MATERIAL			AMPERAGE	VOLTS	TRAVEL SPEED
	CLASS	DIAM.	IPM			
ROOT # 1	E7IT-11	0.068	66	150	18	MANUAL
LAYER # 2 & 3	E7IT-11	0.068	66	150	18	MANUAL
COVER						

BEAD SEQUENCE

PASS 1: 5/16" MAX. SIZE FOR SINGLE-PASS FILLET
PASS 2 AND 3: 3/8" MIN. SIZE FOR MULTIPLE-PASS FILLET

PASS 3
PASS 2
PASS 1
1.

JOINT DETAIL

NOTE:
MATERIAL THICKNESS WILL VARY
DEPENDING ON THE WELD SIZE
QUALIFIED.

Figure 13-27. A fillet weld soundness test determines the parameters for weld size.

Figure 13-28. The PQR test requires cross-examination of fillet welds to determine penetration into the root of the plates and weld size for single- and multiple-pass welds.

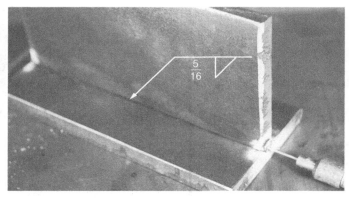

Figure 13-29. The weld is started on one end, stopped in the center, restarted, and then continued to the end.

The weld test for qualification of welders for fillet welds is shown in **Figure 13-29,** using the WPS in **Figure 13-30** and welded in the position required. (Refer to Figure 13-5 for various test positions of fillet welds.) After the weld test is completed, it is inspected for undercut and weld size. Refer to Figure 13-19 and see **Figure 13-31.** The test is cut apart, and the weld joint is polished, **Figure 13-32.** An acid is used to etch the weld joint cross section. The penetration must extend to the root of the joint but not necessarily beyond. This is called a *macroetch test.*

Welder fillet weld tests also require a break test on the center section of the test plate. The test plates are bent with the weld root in tension until the plates are flat or the weld fails, **Figure 13-33.** If the weld does not fracture, the test is satisfactory.

If the weld breaks, the break is examined for complete fusion of the root with no porosity larger than

WELDING PROCEDURE SPECIFICATION

PRODUCER _PCC_
WELDING PROCESS-FCAW _X_
POSITION _HORIZONTAL_

WPS IDENTIFICATION NO. _003_
REVISION NO. _00_ DATE _2-98_
SUPPORTING PQR _003_
REVISION NO. _00_

JOINT DESIGN
V-GROOVE _____
INCLUDED ANGLE _____
ROOT SPACING _____
BACKING MATERIAL _____
TYPE _____ THICKNESS _____

POSITION OF WELD
GROOVE _____
VERTICAL WELDS-UPHILL _____
FILLET _HORIZONTAL_ _____

BASE MATERIALS
SPECIFICATION _ASTM_
TYPE _A 36_
THICKNESS _3/8"_

ELECTRICAL CHARACTERISTICS
DIRECT CURRENT _ELECTRODE NEG._
ELECTRICAL STICKOUT _1/2"-3/4"_

FILLER MATERIAL
AWS SPECIFICATION _AWS A5.20_
MANUFACTURER _LINCOLN_
TRADE NAME _NR232_

TECHNIQUE
STRINGER BEAD _X_
WEAVE BEAD ___ MAXIMUM _____
SINGLE PASS _X_ MULTI PASS _____
INTERPASS CLEANING _WIRE BRUSH_

PREHEAT
PREHEAT TEMP. MIN. _350°F_
INTERPASS TEMPERATURE _—_
MIN _____ MAX _____

POSTHEAT
TEMPERATURE _—_
TIME _—_

WELDING PROCEDURE

LAYER OR PASSES	FILLER MATERIAL			AMPERAGE	VOLTS	TRAVEL SPEED
	CLASS	DIAM.	IPM			
ROOT	E7IT-8	0.068	150	250	20	MANUAL
LAYER						
COVER						

BEAD SEQUENCE	JOINT DETAIL
SINGLE-PASS WELD WELD TO CENTER. STOP, RESTART, AND CONTINUE TO END.	

Figure 13-30. This WPS contains the required information for a welder qualification test for fillet welds.

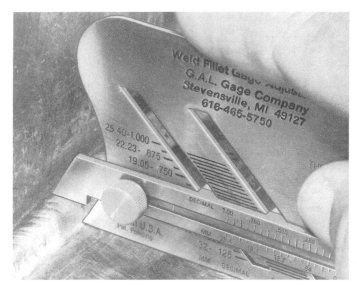

Figure 13-31. The pointer of the weld fillet gage is placed on the edge of the weld leg. The height of the arrow indicates weld leg size. (G.A.L. Gage Co.)

Figure 13-33. The break test is made with the crown facing upward to place the weld in tension as the ram moves downward. Continue the test until the plates are flat or the weld breaks.

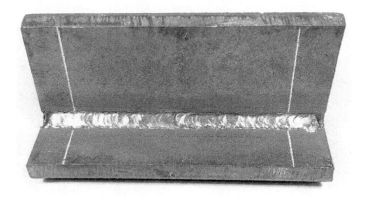

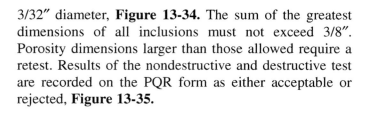

Figure 13-32. A fillet weld test is marked for macroetch test cuts. This operation should be done on a bandsaw rather than using an oxyacetylene cutting torch. Polish and etch for weld size and penetration.

Figure 13-34. The test shown has broken the weld before the plates were flattened.

3/32″ diameter, **Figure 13-34.** The sum of the greatest dimensions of all inclusions must not exceed 3/8″. Porosity dimensions larger than those allowed require a retest. Results of the nondestructive and destructive test are recorded on the PQR form as either acceptable or rejected, **Figure 13-35.**

Rejected Test

If one section of the test fails, the entire test is rejected. A retest may be made under certain conditions:

- An immediate retest may be made with a double test of the failed item.
- A single retest of the failed area may be made after reinstruction.

Welder Qualification and Certification

Upon successful completion of the welding test, a Welder Qualification Test Record is completed. The form is signed by the tester, certifying the statements on the record are correct. See **Figure 13-36.** The terms "qualified welder" and "certified welder" describe the person who has completed an acceptable weld test.

The AWS Code considers a welder to be qualified indefinitely unless:

- The welder is not engaged in a given process of welding for which he or she is qualified for a

PROCEDURE QUALIFICATION RECORD

DATE _2-98_   WPS NO. _003_   REV. NO. _00_   PQR NO. _003_

GUIDED BEND TEST

Specimen Number	Type of Bend	Result	Remarks

PLATE WELD TEST

APPEARANCE _____

UNDERCUT _____

CONVEXITY _____

CONCAVITY _____

WELD SIZE _____

MACRO ETCH _____

BREAK TEST _____

FILLET WELD TEST

SATISFACTORY

SATISFACTORY

SATISFACTORY

SATISFACTORY

SATISFACTORY

1 _O. K._ # 2 _O. K._ # 3 _____

SATISFACTORY

DATE TEST _2-98_   ACCEPTANCE _O.K._   REJECTION _____

PRODUCER _PCC_

TITLE _____

DATE _2-98_

Figure 13-35. The test results are recorded on the PQR test form as either accepted or rejected.

period exceeding six months. (A requalification test may be made on 3/8″ material.)
- There is a specific reason to question the welder's ability.

Tackers, Welding Operators, and Other Qualification Tests

The tests for tackers and welding operators, and other types of welding tests specified in the code are not presented in this text. For information regarding these qualification tests, contact the American Welding Society regarding *AWS D1.1 Structural Welding Code—Steel.*

Review Questions

Please do not write in this text. Write your answers on a separate sheet of paper.
1. What AWS test qualifies welders who demonstrate proficiency in a specific welding process?
2. What two codes govern welding operations?

WELDER QUALIFICATION TEST RECORD

Welder or welding operator's name _____ Identification no. _____

Welding process _____ Semiautomatic _____

Position _____

(Flat, horizontal, overhead or vertical - if vertical, state whether upward or downward)

Welding procedure specification no. _____ Rev. _____ Date _____

Base material specification _____ Backing material _____

Joint thickness _____

Thickness range qualified _____

FILLER METAL

Specification no. _____ Classification _____

Describe filler metal (if not covered by AWS specification) _____

_____

Filler metal diameter and trade name _____

INSPECTION

Appearance _____ Undercut _____ Porosity _____

Guided Bend Test Results

	Type	Result	Remarks
1.			
2.			
3.			
4.			

Fillet Test Results

Appearance _____ Fillet size _____

Fillet weld break test _____ Marcoetch _____

RADIOGRAPHIC TEST RESULTS

Film identification	Results	Remarks	Film identification	Results	Remarks

Test Date _____

Test Approval _____ Test Rejected _____

Inspector _____ Date _____

Figure 13-36. Welder qualification test results are recorded on a certification test form. The test data used for a particular qualification test are indicated on the form. The form will indicate whether the welder passed or failed the test.

3. When is a procedure no longer considered "qualified"?
4. Joints made with materials that have been approved by a governing welding code are called _____.
5. Explain the basis for welder qualification.
6. List three ways to reduce problems associated with weld test failure.
7. What is the rejection criteria for cracks in a weld qualification test?
8. What is the purpose of a fillet weld soundness test?
9. Under what conditions may a retest be performed?
10. What two terms are used to describe a welder who has completed an acceptable weld test?

CHAPTER 14 General Welding Procedures

Objective

After studying this chapter, you will be able to recognize the parameters and variables on the given welding schedules and apply them as a starting point for your own production welding applications.

Purpose of a Welding Procedure

Many welding procedures have been created for various welding tasks. Depending on the experience of the welder or person developing the procedure, and the needs of the application or industry, each procedure is customized and unique.

Electrode manufacturers have developed procedures for each of their products. An electrode will be used with different material thicknesses, applications, and positions. The manufacturer does not know the exact conditions of use. Therefore, a procedure will yield the minimum and maximum settings for electrode speed and voltage requirements. Other considerations include equipment limitations or experience of the welder applying the procedure.

The purpose of using a welding procedure is to obtain a weld quality that meets the specification and to deposit the weld material in a cost-efficient manner. The procedures in this chapter have been developed and used for production welding. You may wish to consider these applications as a starting point for your specific procedure and to improve upon them for your own areas.

Test Welds

Many factors affect the FCAW process. It is important that your procedure be confirmed by test welds before actual welding begins. A test weld and the required weld should have the same joint design, metal, and material thickness. The test weld should be inspected for quality upon completion. The weld must perform its intended function. Testing can be done by visual inspection, cutting welds apart for a macrotest, pulling until destruction, bending the welded joint in a press, applying a penetrant, or performing one of any number of tests.

Producing Good Welds

Once you have started welding, visually check your welds as they are made. The welds should look like your test welds. If they do not, stop and find out why. Minor changes in your schedule can radically affect completed welds. Any changes will impact the quality of each weld. Tolerances should be established for each parameter or variable.

Good welds can be produced by skilled welders using proven schedules within established tolerances. Faulty welds can be produced by the same welders using unproven schedules or tolerances.

FLUX CORED ARC WELDING (FCAW) SCHEDULE

WELD TYPE *FILLET*   POSITION *HORIZONTAL*

BASE METAL TYPE *A36 STEEL*   THICKNESS *3/16"*

ELECTRODE *NR211 MP*   CLASS *E71T-11*   DIA. *0.068*

ELECTRODE IPM *65* AMPS *150*   ELECTRODE STICKOUT *3/4"*

CURRENT TYPE *DCEN*   VOLTAGE (ARC) *17–18*

GAS TYPE —— % ——   GAS TYPE —— % —— CFH ——

NOZZLE TYPE ——   NOZZLE DIAMETER ——

PREHEAT TEMPERATURE MIN. ——   POSTHEAT TEMPERATURE ——

INTERPASS TEMPERATURE ——   TIME ——

MIN. ——   MAX. ——

NOTES AND SPECIAL INSTRUCTIONS:

FLUX CORED ARC WELDING (FCAW) SCHEDULE

WELD TYPE *BUILD-UP/OVERLAY*　　POSITION *FLAT*

BASE METAL TYPE *ALLOY STEEL*　　THICKNESS *ALL*

ELECTRODE *LINCORE 33*　　CLASS — DIA. *5/64"*

ELECTRODE IPM *175* AMPS —　　ELECTRODE STICKOUT *2"*

CURRENT TYPE *DCEP*　　VOLTAGE (ARC) *25–27*

GAS TYPE — % —　　GAS TYPE — % — CFH —

NOZZLE TYPE —　　NOZZLE DIAMETER —

PREHEAT TEMPERATURE MIN. *300°F*　　POSTHEAT TEMPERATURE *600°F*

INTERPASS TEMPERATURE _____　　TIME *15 MIN.*

MIN. *300°F*　　MAX. *400°F*

NOTES AND SPECIAL INSTRUCTIONS:

FLUX CORED ARC WELDING (FCAW) SCHEDULE

WELD TYPE _OVERLAY_   POSITION _FLAT HORIZONTAL_

BASE METAL TYPE _STEEL_   THICKNESS —

ELECTRODE _LINCORE 50_   CLASS —   DIA. _7/64"_

ELECTRODE IPM _100_ AMPS _375_   ELECTRODE STICKOUT _1 1/4"_

CURRENT TYPE _DCEP_   VOLTAGE (ARC)

GAS TYPE — % —   GAS TYPE — % — CFH —

NOZZLE TYPE —   NOZZLE DIAMETER —

PREHEAT TEMPERATURE MIN. —   POSTHEAT TEMPERATURE —

INTERPASS TEMPERATURE —   TIME —

MIN. —   MAX. —

NOTES AND SPECIAL INSTRUCTIONS:

FLUX CORED ARC WELDING (FCAW) SCHEDULE

WELD TYPE _SURFACE OVERLAY_   POSITION _FLAT_

BASE METAL TYPE _A36 STEEL_   THICKNESS _1/4"_

ELECTRODE _NR211 MP_   CLASS _E71T-11_   DIA. _0.068"_

ELECTRODE IPM _52_   AMPS _125_   ELECTRODE STICKOUT _3/4"_

CURRENT TYPE _DCEN_   VOLTAGE (ARC) _18–19_

GAS TYPE — % —   GAS TYPE — % — CFH —

NOZZLE TYPE —   NOZZLE DIAMETER —

PREHEAT TEMPERATURE MIN. —   POSTHEAT TEMPERATURE —

INTERPASS TEMPERATURE —   TIME —

MIN. —   MAX. —

NOTES AND SPECIAL INSTRUCTIONS:

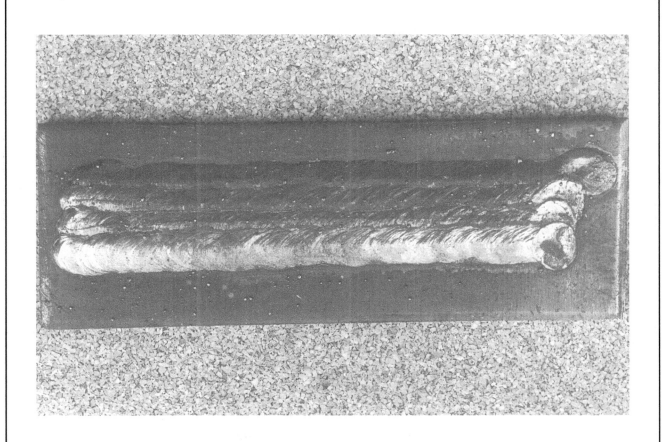

FLUX CORED ARC WELDING (FCAW) SCHEDULE

WELD TYPE *V-GROOVE*   POSITION *UPHILL*

BASE METAL TYPE *A36 STEEL*   THICKNESS *3/16"*

ELECTRODE *NR211 MP*   CLASS *E71T-11*   DIA. *0.068"*

ELECTRODE IPM *40*   AMPS *110*   ELECTRODE STICKOUT *1/2"–3/4"*

CURRENT TYPE *DCEN*   VOLTAGE (ARC) *15–16*

GAS TYPE —   % —   GAS TYPE —   % — CFH —

NOZZLE TYPE —   NOZZLE DIAMETER —

PREHEAT TEMPERATURE MIN. —   POSTHEAT TEMPERATURE —

INTERPASS TEMPERATURE —   TIME —

MIN. —   MAX. —

NOTES AND SPECIAL INSTRUCTIONS:

FLUX CORED ARC WELDING (FCAW) SCHEDULE

WELD TYPE _FILLET_   POSITION _HORIZONTAL_

BASE METAL TYPE _A36 STEEL_   THICKNESS _3/8"_

ELECTRODE _NR211 MP_   CLASS _E71T-11_   DIA. _0.068"_

ELECTRODE IPM _75_   AMPS _180_   ELECTRODE STICKOUT _1/2"–1"_

CURRENT TYPE _DCEN_   VOLTAGE (ARC) _18–20_

GAS TYPE —   % —   GAS TYPE —   % — CFH —

NOZZLE TYPE —   NOZZLE DIAMETER —

PREHEAT TEMPERATURE MIN. —   POSTHEAT TEMPERATURE —

INTERPASS TEMPERATURE —   TIME —

MIN. —   MAX. —

NOTES AND SPECIAL INSTRUCTIONS:

FLUX CORED ARC WELDING (FCAW) SCHEDULE

WELD TYPE *GROOVE/FILLET*   POSITION *FLAT/HORIZONTAL*

BASE METAL TYPE *STEEL*   THICKNESS *1/8"*

ELECTRODE *NR211 MP*   CLASS *E7IT-11*   DIA. *0.035"*

ELECTRODE IPM *155* AMPS *120*   ELECTRODE STICKOUT *1/2"–3/4"*

CURRENT TYPE *DCEN*   VOLTAGE (ARC) *17*

GAS TYPE — % —   GAS TYPE — % — CFH —

NOZZLE TYPE —   NOZZLE DIAMETER —

PREHEAT TEMPERATURE MIN. —   POSTHEAT TEMPERATURE —

INTERPASS TEMPERATURE —   TIME —

MIN. —   MAX. —

NOTES AND SPECIAL INSTRUCTIONS:

REMOVE ALL OIL, GREASE, AND RUST BEFORE WELDING.

FLUX CORED ARC WELDING (FCAW) SCHEDULE

WELD TYPE *FILLET*   POSITION *HORIZONTAL*

BASE METAL TYPE *HOT ROLLED STEEL*   THICKNESS *1/16"*

ELECTRODE *COREX SELF SHIELD*   CLASS ——   DIA. *0.030"*

ELECTRODE IPM —— AMPS *50*   ELECTRODE STICKOUT *1/2"*

CURRENT TYPE *DCEN*   VOLTAGE (ARC) ——

GAS TYPE —— % ——   GAS TYPE —— % —— CFH ——

NOZZLE TYPE ——   NOZZLE DIAMETER ——

PREHEAT TEMPERATURE MIN. ——   POSTHEAT TEMPERATURE ——

INTERPASS TEMPERATURE ——   TIME ——

MIN. ——   MAX. ——

NOTES AND SPECIAL INSTRUCTIONS:

FLUX CORED ARC WELDING (FCAW) SCHEDULE

WELD TYPE *FILLET* POSITION *HORIZONTAL*

BASE METAL TYPE *A36 STEEL* THICKNESS *3/16"–1/4"*

ELECTRODE *7100* CLASS — DIA. *1/16"*

ELECTRODE IPM *138* AMPS *200* ELECTRODE STICKOUT *3/4"*

CURRENT TYPE *DCEP* VOLTAGE (ARC) *20*

GAS TYPE CO_2 % *100* GAS TYPE — % — CFH *35–40*

NOZZLE TYPE — NOZZLE DIAMETER *$\frac{5"}{8} - \frac{3"}{4}$*

PREHEAT TEMPERATURE MIN. — POSTHEAT TEMPERATURE —

INTERPASS TEMPERATURE — TIME —

MIN. — MAX. —

NOTES AND SPECIAL INSTRUCTIONS:

FLUX CORED ARC WELDING (FCAW) SCHEDULE

WELD TYPE *V-GROOVE*   POSITION *DOWNHILL 45°*

BASE METAL TYPE *A36 STEEL*   THICKNESS *3/16"*

ELECTRODE *NR211 MP*   CLASS *E7IT-11*   DIA. *0.068"*

ELECTRODE IPM *66*   AMPS *150*   ELECTRODE STICKOUT *3/4"*

CURRENT TYPE *DCEN*   VOLTAGE (ARC) *18–19*

GAS TYPE —   % —   GAS TYPE —   % — CFH —

NOZZLE TYPE —   NOZZLE DIAMETER —

PREHEAT TEMPERATURE MIN. —   POSTHEAT TEMPERATURE —

INTERPASS TEMPERATURE —   TIME —

MIN. —   MAX. —

NOTES AND SPECIAL INSTRUCTIONS:

FLUX CORED ARC WELDING (FCAW) SCHEDULE

WELD TYPE _FILLET_   POSITION _UPHILL_

BASE METAL TYPE _A36 STEEL_   THICKNESS _5/8"_

ELECTRODE _NR211 MP_   CLASS _E71T-11_   DIA. _0.068"_

ELECTRODE IPM _63_ AMPS _150_   ELECTRODE STICKOUT _3/4"_

CURRENT TYPE _DCEN_   VOLTAGE (ARC) _17–18_

GAS TYPE — % —   GAS TYPE — % — CFH —

NOZZLE TYPE —   NOZZLE DIAMETER —

PREHEAT TEMPERATURE MIN. —   POSTHEAT TEMPERATURE —

INTERPASS TEMPERATURE —   TIME —

MIN. —   MAX. —

NOTES AND SPECIAL INSTRUCTIONS:

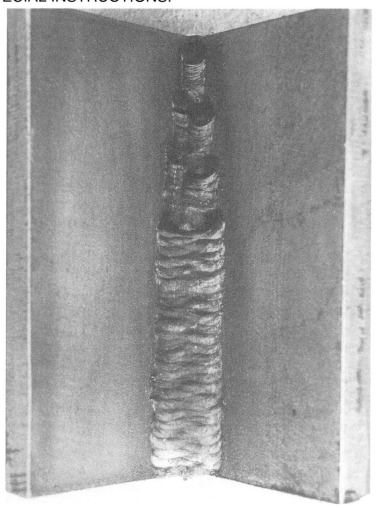

FLUX CORED ARC WELDING (FCAW) SCHEDULE

WELD TYPE _FILLET_   POSITION _HORIZONTAL_

BASE METAL TYPE _A36 STEEL_   THICKNESS _1/4"–3/8"_

ELECTRODE _NR211 MP_   CLASS _E71T-11_   DIA. _0.068"_

ELECTRODE IPM _75_   AMPS _180_   ELECTRODE STICKOUT _3/4"–1"_

CURRENT TYPE _DCEN_   VOLTAGE (ARC) _18–19_

GAS TYPE _—_   % _—_   GAS TYPE _—_   % _—_ CFH _—_

NOZZLE TYPE _—_   NOZZLE DIAMETER _—_

PREHEAT TEMPERATURE MIN. _—_   POSTHEAT TEMPERATURE _—_

INTERPASS TEMPERATURE _—_   TIME _—_

MIN. _—_   MAX. _—_

NOTES AND SPECIAL INSTRUCTIONS:

FLUX CORED ARC WELDING (FCAW) SCHEDULE

WELD TYPE *V-GROOVE*   POSITION *FLAT*

BASE METAL TYPE *A36 STEEL*   THICKNESS *1"*

ELECTRODE *NR232*   CLASS *E71T-8*   DIA. *0.068"*

ELECTRODE IPM *110* AMPS *200*   ELECTRODE STICKOUT *3/4"*

CURRENT TYPE *DCEN*   VOLTAGE (ARC) *18-20*

GAS TYPE —— % ——   GAS TYPE —— % —— CFH ——

NOZZLE TYPE ——   NOZZLE DIAMETER ——

PREHEAT TEMPERATURE MIN. *350°F*   POSTHEAT TEMPERATURE ——

INTERPASS TEMPERATURE *350°F*   TIME ——

MIN. *350°F*   MAX. *400°F*

NOTES AND SPECIAL INSTRUCTIONS:

REMOVE ALL RUST, SCALE, AND SLAG BEFORE WELDING.

FLUX CORED ARC WELDING (FCAW) SCHEDULE

WELD TYPE _FILLET_   POSITION _HORIZONTAL_

BASE METAL TYPE _A36 STEEL_   THICKNESS _3/8"_

ELECTRODE _NR203 Ni_   CLASS _—_   DIA. _3/32"_

ELECTRODE IPM _84_ AMPS _260_   ELECTRODE STICKOUT _3/4"_

CURRENT TYPE _DCSP_   VOLTAGE (ARC) _21_

GAS TYPE _—_   % _—_   GAS TYPE _—_ % _—_ CFH _—_

NOZZLE TYPE _—_   NOZZLE DIAMETER _—_

PREHEAT TEMPERATURE MIN. _—_   POSTHEAT TEMPERATURE _—_

INTERPASS TEMPERATURE _—_   TIME _—_

MIN. _—_   MAX. _—_

NOTES AND SPECIAL INSTRUCTIONS:

FLUX CORED ARC WELDING (FCAW) SCHEDULE

WELD TYPE *FILLET*   POSITION *ALL*

BASE METAL TYPE *A36 STEEL*   THICKNESS *1/4"–3/8"*

ELECTRODE *NR211 MP*   CLASS *E71T-11*   DIA. *5/64"*

ELECTRODE IPM *60*   AMPS *175*   ELECTRODE STICKOUT *3/4"*

CURRENT TYPE *DCEN*   VOLTAGE (ARC) *16–17*

GAS TYPE —   % —   GAS TYPE — % — CFH —

NOZZLE TYPE —   NOZZLE DIAMETER —

PREHEAT TEMPERATURE MIN. —   POSTHEAT TEMPERATURE —

INTERPASS TEMPERATURE —   TIME —

MIN. —   MAX. —

NOTES AND SPECIAL INSTRUCTIONS:

FLUX CORED ARC WELDING (FCAW) SCHEDULE

WELD TYPE *FILLET*   POSITION *HORIZONTAL*

BASE METAL TYPE *A36 STEEL*   THICKNESS —

ELECTRODE *NR203 Ni*   CLASS —   DIA. *5/64"*

ELECTRODE IPM *105* AMPS *250*   ELECTRODE STICKOUT *3/4"*

CURRENT TYPE *DCEN*   VOLTAGE (ARC) *20–21*

GAS TYPE —   % —   GAS TYPE —   % — CFH —

NOZZLE TYPE —   NOZZLE DIAMETER —

PREHEAT TEMPERATURE MIN. —   POSTHEAT TEMPERATURE —

INTERPASS TEMPERATURE —   TIME —

MIN. —   MAX. —

NOTES AND SPECIAL INSTRUCTIONS:

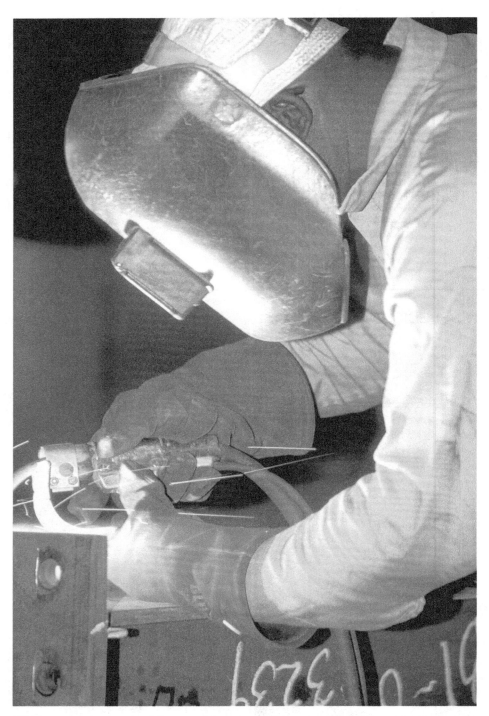

Welders must closely follow welding procedures to produce welds that meet specifications.

Reference Section

The following pages contain over 20 charts that will be useful as reference in a variety of welding-related areas. To make locating information easier, the charts are listed below by title and page number.

	Guide for Shade Numbers			
Operation	Electrode Size 1/32 in. (mm)	Arc Current (A)	Minimum Protective Shade	Suggested[1] Shade No. (Comfort)
Shielded metal arc welding	Less than 3 (2.5) 3–5 (2.5–4) 5–8 (4–6.4) More than 8 (6.4)	Less than 60 60–160 160–250 250–550	7 8 10 11	— 10 12 14
Gas metal arc welding and flux cored arc welding		Less than 60 60–160 160–250 250–500	7 10 10 10	— 11 12 14
Gas tungsten arc welding		Less than 50 50–150 150–500	8 8 10	10 12 14
Air carbon arc cutting	(Light) (Heavy)	Less than 500 500–1000	10 11	12 14
Plasma arc welding		Less than 20 20–100 100–400 400–800	6 8 10 11	6 to 8 10 12 14
Plasma arc cutting	(Light)[2] (Medium)[2] (Heavy)[2]	Less than 300 300–400 400–800	8 9 10	9 12 14
Torch brazing		—	—	3 or 4
Torch soldering		—	—	2
Carbon arc welding		—	—	14
	Plate thickness			
	in.	mm		
Gas welding Light Medium Heavy	Under 1/8 1/8 to 1/2 Over 1/2	Under 3.2 3.2 to 12.7 Over 12.7		4 or 5 5 or 6 6 or 8
Oxygen cutting Light Medium Heavy	Under 1 1 to 6 Over 6	Under 25 25 to 150 Over 150		3 or 4 4 or 5 5 or 6

[1] As a rule of thumb, start with a shade that is too dark to see the weld zone. Then go to a lighter shade which gives sufficient view of the weld zone without going below the minimum. In oxyfuel gas welding or cutting where the torch produces a high yellow light, it is desirable to use a filter lens that absorbs the yellow or sodium line in the visible light of the (spectrum) operation.

[2] These values apply where the actual arc is clearly seen. Experience has shown that lighter filters may be used when the arc is hidden by the workpiece.

Chart R-1. Guide for shade numbers.

Electrode	Charpy V-notch, ft./lb (°F)	Shielding	Current	Single/ multipass	Applications
Carbon and Low-Alloy Steel Electrodes for FCAW per AWS A5.20					
E70T-1	20(0)	CO_2	DCEP	multi	General-purpose flat and horizontal welding. Railcar fabrication, beams and girders, ships, over-the-road vehicles, storage tanks.
E70T-2	—	CO_2	DCEP	single	For single-pass welds on rusted, contaminated base material. Castings, machine bases, oil-field equipment, railcars.
E70T-3	—	self	DCEP	single	Rapid, automated welding on thin-gage steel.
E70T-4	—	self	DCEP	multi	
E70T-5	20 (−20)	CO_2	DCEP	multi	Basic slag for good impact properties, low weldmetal hydrogen, low crack sensitivity.
E70T-6	20 (−20)	self	DCEP	multi	
E70T-7	—	self	DCEN	multi	
E70T-10	—	self	DCEN	multi	
E70T-G	—	b.	b.	b.	
E71T-1	20(0)	CO_2	DCEP	multi	General-purpose all-position welding. Use at low currents (150–200 A) to bridge wide gaps, at higher currents (200–250 A) for DCEP penetration. Barges, oil rigs, storage vessels, earthmoving equipment.
E71T-5	20 (−20)	CO_2	DCEP	multi	
E71T-7	—	self	DCEN	multi	
E71T-8	20 (−20)	self	DCEN	multi	
E71T-11	—	self	DCEN	multi	Thin-gage material, structural steel
E71T-GS	—	a.	b.	b.	

a. Argon additions may improve weldmetal properties and welding characteristics.
b. New or proprietary wire, properties as specified by the supplier.

Various Electrode Sizes for FCAW		
Thousandths/ inches	**Fractional/ inches**	**Metric/ millimeter**
0.030 0.035		0.8 0.9
0.045 0.052	3/64	1.2 1.3
0.062 0.068	1/16	1.6
0.072 0.078	5/64	1.98
0.093 0.109	3/32 7/64	2.4 2.8
0.120 0.128	1/8	3.2
	5/32 3/16	4.0 4.8

Chart R-3. Nominal sizes for FCAW electrodes.

FCAW Steel Electrodes	
Diameter	**Inches per Pound***
0.035″ (0.9 mm)	4,350
0.045″ (1.2 mm)	2,500
1/16″ (1.6 mm)	1,300
5/64″ (2.0 mm)	925
3/32″ (2.4 mm)	615
7/64″ (2.8 mm)	550
0.120″ (3.0 mm)	420

***Approximate**—Values will vary with AWS class and wire type.

Chart R-4. FCAW steel electrodes.

Typical Operating Parameters for Carbon Steels (without Shielding Gas)			
Wire Diameter		**Amperage**	**Voltage**
0.035" 0.035" Optimum			
0.045" 0.045" Optimum		95–180 130	15–17 15
0.052" 0.052" Optimum			
1/16" 1/16" Optimum		100–300 200	18–20 19
0.068" 0.068" Optimum		120–275 190	17–22 18
0.072" 0.072" Optimum		130–250 255	16–25 20–23
5/64" 5/64" Optimum		150–300 235	21 20
3/32" 3/32" Optimum		245–400 305	24 21
7/64" 7/64" Optimum		325–625 450	23–25 27
0.120" 0.120" Optimum		400–550 450	28–31 29

Notes:
1. 0.35" Dia. ES/O   1/2"
2. 0.45" Dia. ES/O   3/4"
3. 1/16" Dia. and up ES/O   3/4"
4. 0.120" Dia. ES/O   2 3/4"
5. Use lower end of range for thinner materials.
6. Use middle of range for vertical welding.
7. Use upper range for flat and horizontal welding only.
8. Use DCRP-DCSP as specified by electrode manufacturer.

Chart R-5. Operating parameters for carbon steels without shielding gas.

Typical Operating Parameters for Carbon Steels (with Shielding Gas)				
Wire Diameter		**Amperage**	**Voltage**	**Shielding**
0.035" 0.035" Optimum		130–280 200	20–30 27	CO_2 or 75/25
0.045" 0.045" Optimum		150–290 240	23–30 26	CO_2 or 75/25
0.052" 0.052" Optimum		180–310 260	24–32 26	CO_2 or 75/25
1/16" 1/16" Optimum		180–400 275	25–34 28	CO_2 or 75/25
5/64" 5/64" Optimum		250–400 350	26–30 27	CO_2 or 75/25
3/32" 3/32" Optimum		350–550 450	26–33 31	CO_2 or 75/25
7/64" 7/64" Optimum		500–700 625	30–35 34	CO_2 or 75/25
1/8" 1/8" Optimum		600–800 725	31–36 34	CO_2 or 75/25

Notes:
1. Use standard CV welding machine.
2. All procedures use DCRP.
3. Electrical stickout:
    0.035"–1/16" diameter electrodes 3/8" to 3/4".
    5/64"–1/8" diameter electrodes 3/4" to 1 1/4".
4. Voltages may be reduced 1 1/2 V for 75/25 mixture.
5. Use lower end of range for thinner materials.
6. Use lower end of range for uphill welding.
7. Use middle of range for overhead position welding.
8. Use optimum parameters for flat position welding.
9. Use 5/64"–3/32"–7/64"–1/8" for flat only.

Chart R-6. Operating parameters for carbon steels with shielding gas.

55555

Stainless Steel Flux Cored Electrodes

AWS A5.22 Standard Product Alloy	Variation Product Alloy	Flat Uphill 0.035"	0.045"	0.063"	Out of Position 0.045"	High Deposition 0.045"	0.063"	Applications
E308 T-1		•	•	•				Most often used to weld base metal of similar analysis such as AISI 301, 302, 304, 305, and 308.
E308L T-1		•	•	•	•	•	•	Low carbon content improves intergranular corrosion resistance over E308 type. General applications use.
	*308LA T-1		•	•				Primarily used for equipment & piping carrying liquefied helium, nitrogen, and natural gas. Good impact properties at cryogenic temperatures.
	308LN T-1		•					Used to weld nitrogen alloy AISI 304LN.
E309 T-1		•	•	•				Used to weld similar alloys in wrought or cast form and AISI 304 in severe corrosion conditions. Also used to weld AISI 304 to carbon steel.
	309Mo T-1	•	•	•				The addition of Mo to E309 allows use as a first layer cladding on carbon steel which will have further overlays. Also used for joining CrNi and CrNiMo stainless steels to each other and to carbon steel.
E309L T-1		•	•	•	•	•	•	Low carbon content gives better resistance to intergranular corrosion than does E309 T-1.
	309MoL T-1	•	•	•	•	•	•	Typically used to join dissimilar CrNi and CrNiMo stainless steels to each other and to carbon steel and as first layer on carbon steel to be clad with stainless overlay. Low carbon content improves crack resistance.
E309CbL T-1			•					Used to overlay carbon and low alloy steels and to produce a columbium-stabilized first layer.
E310 T-1			•					Most often used to weld AISI 310, 310S, and similar alloys.
E316 T-1		•	•	•				Typically used for welding similar alloys. The presence of molybdenum provides increased creep resistance at elevated temperatures.
E316L T-1		•	•	•	•	•	•	Low carbon increases the resistance to intergranular corrosion but decreases the elevated temperature properties. Commonly used to join AISI 304, 304L, 316, 316L.
	316CuL T-1		•	•				The addition of copper to E316LT-I provides corrosion resistance in higher concentrations of sulfuric acid.
E317L T-1		•	•	•				Used for welding alloys of similar composition and are usually limited to severe corrosion applications involving sulfuric and sulfurous acids and their salts.
	318L T-1		•					For welding AISI 316Cb and 316Ti alloys.
E347 T-1			•	•				This columbium stabilized alloy is generally used to weld AISI 321 and 347 materials.
	*347L T-1	•	•	•				The low carbon content provides better resistance to intergranular corrosion than does E347T-I. Used for final pass overlay on AISI 321 and 347 materials.
E410 T-G			•					Most commonly used to weld alloys of similar analysis. Also used for surfacing of carbon steels to resist corrosion, erosion, and abrasion.
	2209 T-1		•					For welding Avesta Sheffield 2205 Code Plus Two® and similar alloys, to each other, to carbon steels, and to stainless steels.
	NiCr-3 T-1		•					For welding Alloy 600, Alloy 800 and similar materials to each other, to carbon steels, and to stainless steels.
	NiCrMo-3 T-1		•					For welding highly corrosion-resistant materials such as Alloy 625, Alloy 825, Avesta Sheffield 254 SMO® AL6XN®, and similar 6% Mo type alloys. Also suitable for 317LMn and 9% Ni steel.

*Indicates the variation product falls within the AWS specification for the similar standard product.

Avesta flux cored electrodes are available in a wide selection of alloys and in three usage types. Designed for use with 100% CO_2 or a 75% Argon 25% CO_2 shielding gas, they offer excellent economy in all applications. All alloys and sizes are formulated for flat and uphill use with some out-of-position capability. Some alloys and sizes are also available in all position (AP) and high deposition (HD) formats which provide for even greater economies of use.

Chart R-7. Stainless steel flux cored electrodes. (Avesta Welding Products, Inc.)

Typical Operating Parameters for Stainless Steels (without Shielding Gas)			
Wire Diameter	Amperage	Voltage	Stickout
0.045″	100–180	26–32	3/8″–5/8″
1/16″	150–275	26–32	1/2″–1″
5/64″	200–300	26–32	3/4″–1 1/4″
3/32″	225–350	26–30	1″–1 1/2″
Note: All procedures use DCRP.			

Chart R-8. Operating parameters for stainless steels without shielding gas.

Typical Operating Parameters for Stainless Steels (with Shielding Gas)				
Wire Diameter		Amperage	Voltage	Shielding
0.035″ 0.035″	Optimum	60–150 110	23–29 26	CO_2 or 75/25 75/25
0.045″ 0.045″	Optimum	85–225 160	23–32 27	CO_2 or 75/25 75/25
1/16″ 1/16″	Optimum	160–330 250	25–36 30	CO_2 or 75/25 75/25

Notes:
1. Use standard CV welding machine.
2. All procedures use DCRP.
3. Maintain 1/2″ to 3/4″ stickout.
4. Use 1/2″ stickout for overhead position.
5. Ar 75%–CO_2 25% gas mixtures will have lower spatter.
6. Use backhand technique for all positions.
7. Shielding gas flow rates 25–50 cfh.

Chart R-9. Operating parameters for stainless steels with shielding gas.

Etching Reagents for Microscopic Examination of Iron and Steel			
Application	Etching	Composition	Remarks
Carbon low-alloy and medium-alloy steels	Nital	Nitric acid (sp gr 1.42) .... 1–5 ml Ethyl or methyl alcohol .. 95–99 ml	Darkens perlite and gives contrast between adjacent colonies; reveals ferrite boundaries; differentiates ferrite from martensite; shows case depth of nitrided steel. Etching time: 5–60 secs.
	Picral	Picric acid .... 4 g Methyl alcohol .... 100 ml	Used for annealed and quench-hardened carbon and alloy steel. Not as effective as No. 1 for revealing ferrite grain boundaries. Etching time: 5–120 secs.
	Hydrochloric and picral acids	Hydrochloric acid .... 5 ml Picric acid .... 1 g Methyl alcohol .... 100 ml	Reveals austenitic grain size in both quenched and quenched-tempered-steels.
Alloy and stainless steel	Mixed acids	Nitric acid .... 10 ml Hydrochloric acid .... 20 ml Glycerol .... 20 ml Hydrogen peroxide .... 10 ml	Iron-chromium-nickel-manganese alloy steel. Etching: Use fresh acid.
	Ferric chloride	Ferric chloride .... 5 g Hydrochloric acid .... 20 ml Water, distilled .... 100 ml	Reveals structure of stainless and austenitic nickel steels.
	Marble's reagent	Cupric sulfate .... 4 g Hydrochloric acid .... 20 ml Water, distilled .... 20 ml	Reveals structure of various stainless steels.
High-speed steels	Snyder-Graff	Hydrochloric acid .... 9 ml Nitric acid .... 9 ml Methyl alcohol .... 100 ml	Reveals grain size of quenched and tempered high-speed steels. Etching time: 15 secs. to 5 min.

Chart R-10. Etching reagents.

Etching Procedures—Stainless Steels			
Reagents	**Composition**	**Procedure**	**Uses**
Nitric and acetic acids	HNO_3 . . . . 30 ml Acetic acid . . . . 20 ml	Apply by swabbing.	Stainless alloys and others high in nickel or cobalt.
Cupric sulphate	$CuSO_4$ . . . . 4 gms HCl . . . . 20 ml H_2O . . . . 20 ml	Etch by immersion.	Structure of stainless steels.
Cupric chloride and hydrochloric acid	$CuCl_2$ . . . . 5 gms HCl . . . . 100 ml Ethyl alcohol . . . . 100 ml H_2O . . . . 100 ml	Use cold immersion or swabbing.	Austenitic and ferritic steels.

Chart R-11. Etching procedures for stainless steels.

Etching Procedures—Iron and Steel			
Reagents	**Composition**	**Procedure**	**Uses**
Macro Examination			
Nitric acid	HNO_3 . . . . 5 ml H_2O . . . . 95 ml	Immerse 30 to 60 seconds.	Shows structure of welds.
Ammonium persulphate	$(NH_4)_2S_2O_3$ . . . . 10 gms H_2O . . . . 90 ml	Surface should be rubbed with cotton during etching.	Brings out grain structure, recrystallization at welds.
Nital	HNO_3 . . . . 5 ml Ethyl alcohol . . . . 95 ml	Etch 5 min. followed by 1 sec. in HCl (10%).	Shows cleanness, depth of hardening, carburized or decarburized surfaces, etc.
Micro Examination			
Picric acid (picral)	Picric acid . . . . 4 gms Ethyl or methyl alcohol (95%) . . . . 100 ml	Etching time a few seconds to a minute or more.	For all grades of carbon steels.

Chart R-12. Etching procedures for iron and steel.

			Preheat (°F) (a)					Preheat (°F) (a)
Steel Group	Steel Designation		Carbon	Base Metal 4″ thick	Steel Group	Steel Designation		Carbon	Base Metal 4″ thick

Preheat Recommendation Chart

Steel Group	Steel Designation		Carbon	Base Metal 4″ thick	Steel Group	Steel Designation		Carbon	Base Metal 4″ thick
Carbon steels	AISI-SAE (c)	1015	0.13–0.18	150°	Chromium steels	AISI-SAE	5015	0.12–0.17	200°
		1020	0.18–0.23	150°			5046	0.43–0.48	450°
		1030	0.28–0.34	200°			5115	0.13–0.18	200°
		1040	0.37–0.44	300°			5145	0.43–0.48	450°
		1080	0.75–0.88	600°			5160	0.56–0.64	550°
Manganese steels	AISI-SAE	1330	0.28–0.33	250°	Austenitic manganese and chrome-nickel steels	ASTM	11%–14% Mn	0.5–1.3	Preheat only to remove chill from base metal.
		1335	0.33–0.38	300°			302	0.15 Max.	
		1340	0.38–0.43	350°			309	0.20 Max.	
		1345	0.43–0.48	400°			310	0.25 Max.	
		1345H	0.42–0.49	400°			347	0.08 Max.	(b)
Molybdenum steels	AISI-SAE	4027H	0.24–0.30	250°	Carbon steel plate structural quality	ASTM	A36	0.27 Max.	250°
		4032H	0.29–0.35	300°			A131 Gr. B	0.21 Max.	200°
		4037H	0.34–0.41	350°			A284 Gr. C	0.29 Max.	250°
		4042H	0.39–0.46	400°			(d) A678 Gr. B	0.20 Max.	200°
		4047H	0.44–0.51	450°	High-strength low-alloy steels structural quality	ASTM	A131-H.S.	0.18 Max.	350°
Chrome molybdenum steels	AISI-SAE	4118	0.17–0.23	250°			A242 Type 2	0.20 Max.	200°
		4130	0.27–0.34	300°			A441	0.22 Max.	200°
		4135	0.32–0.39	400°			A588 Gr. B	0.20 Max.	300°
		4145	0.41–0.49	500°			A633 Gr. E	0.22 Max.	250°
		4145H	0.42–0.49	500°	Alloy and pressure vessel quality steels	ASTM	(d) A514 Gr. F	0.10–0.21	350°
Ni-chrome molybdenum and Ni-moly. steels	AISI-SAE	4340	0.38–0.43	500°			(d) A514 Gr. H	0.12–0.21	300°
		4615	0.18–0.18	250°			(d) A514 Gr. Q	0.14–0.21	550°
		4620	0.17–0.22	250°			A515 Gr. 70	0.35 Max.	300°
		4720H	0.17–0.23	300°			A516 Gr. 70	0.30 Max.	250°
		4820H	0.17–0.23	300°					

Notes:
a. These suggested preheats are recommended when low-hydrogen processes are used on base metals that are 4″ thick. Lower preheats would be needed on thinner material while higher preheats would be necessary on thicker materials. When using non-low-hydrogen processes, increase suggested preheats by 300°F.
The steels shown on the chart are only partially representative of the steels used in the manufacture of earthmoving and other machinery. A preheat calculator (WC-8) available from Lincoln Electric Co. makes it possible to figure suggested preheats for other steels based upon the chemistry of the steel and the thickness of the parts to be surfaced.
b. It is sometimes advisable to preheat large, thick 11% to 14% manganese parts before welding. Use a maximum of 200°F preheat. (Do not exceed 500°F preheat and/or interpass temperature.) Check base metal with magnet. 11%–14% manganese and the ASTM 300 series of chrome-nickel steels are *not* magnetic.
c. Low-carbon steel.
d. Q & T Steels—see "Need for Preheat."

Chart R-13. Preheat recommendation chart. (Lincoln Electric Co.)

Preheat, Interpass, and Postheat Temperatures				
AWS Classification[a]	**Preheat and Interpass Temperature[b]**		**PWHT Temperature**	
	°F	°C	°F	°C
E70T5-A1 E80T1-A1 E81T1-A1 E80T5-Nil E80T5-Ni2[c] E80T5-Ni3[c] E90T5-N3[c] E90T5-D2 D100T5-D2	300 ± 25	150 ± 15	1150 ± 25	620 ± 15
E81T1-B1 E80T1-B2 E81T1-B2 E80T5-B2 E80T1-B2H E80T5-B2L E90T1-B3 E91T1-B3 E90T5-B3 E100T1-B3 E90T1-B3H E90T1-B3L	350 ± 25	176 ± 15	1275 ± 25	690 ± 15
E71T8-Ni1 E80T1-Ni1 E81T1-Ni1 E71T8-Ni2 E80T1-Ni2 E81T1-Ni2 E90T1-Ni2 E91T1-Ni2 E91T1-D1 E90T1-D3 E80T5-K1 E70T4-K2 E71T8-K2 E80T1-K2 E80T5-K2 E90T1-K2 E91T1-K2 E90T5-K2 E100T1-K3 E100T5-K3 E110T1-K3 E110T5-K3 E111T1-K4 E110T5-K4 E120T5-K4 E120T1-K5 E61T8-K6 E71T8-K6 E80T1-W	300 ± 25	150 ± 15	None	None
EXXXTX-G	Conditions as agreed upon between supplier and purchaser.			

Notes:

a. In this table, the digit (i.e., 0 or 1) before the letter 'T' indicates the primary welding position for which the electrode is designed (usability). Refer to Figure 5-2.

b. These temperatures are specified for testing under this specification and are not to be considered as recommendations for preheat and postweld heat treatment in production welding. The requirements for production welding must be determined by the user.

c. Postweld heat treatment temperatures in excess of 1150°F (620°C) will decrease the impact value.

Chart R-14. Preheat, interpass, and postheat temperatures.

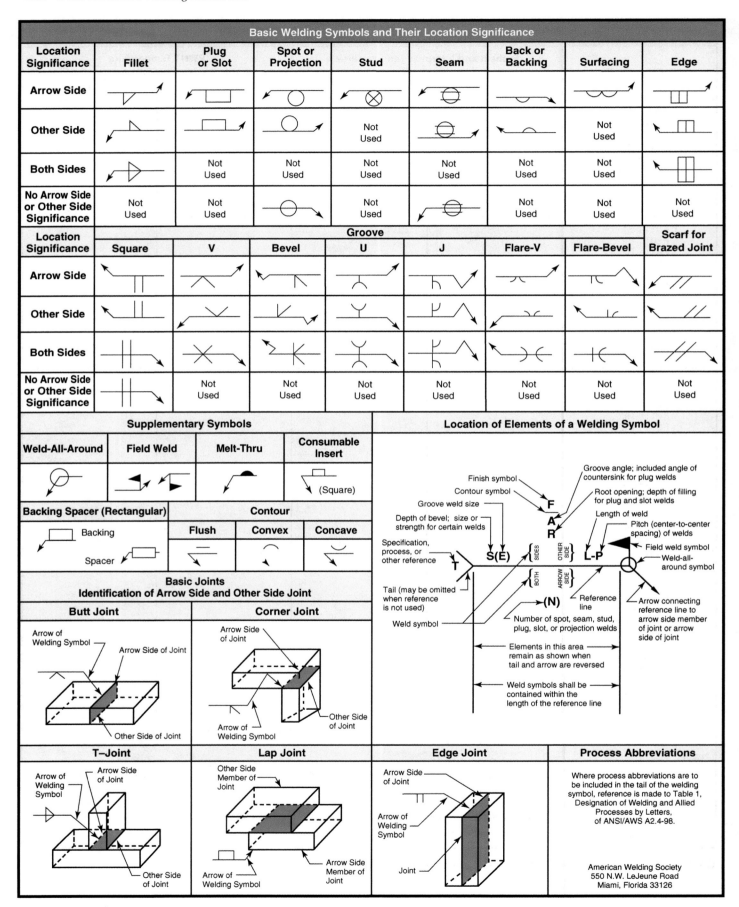

Chart R-15. Weld symbols. (From *ANSI/AWS A2.4-98.* Printed with Permission of the American Welding Society)

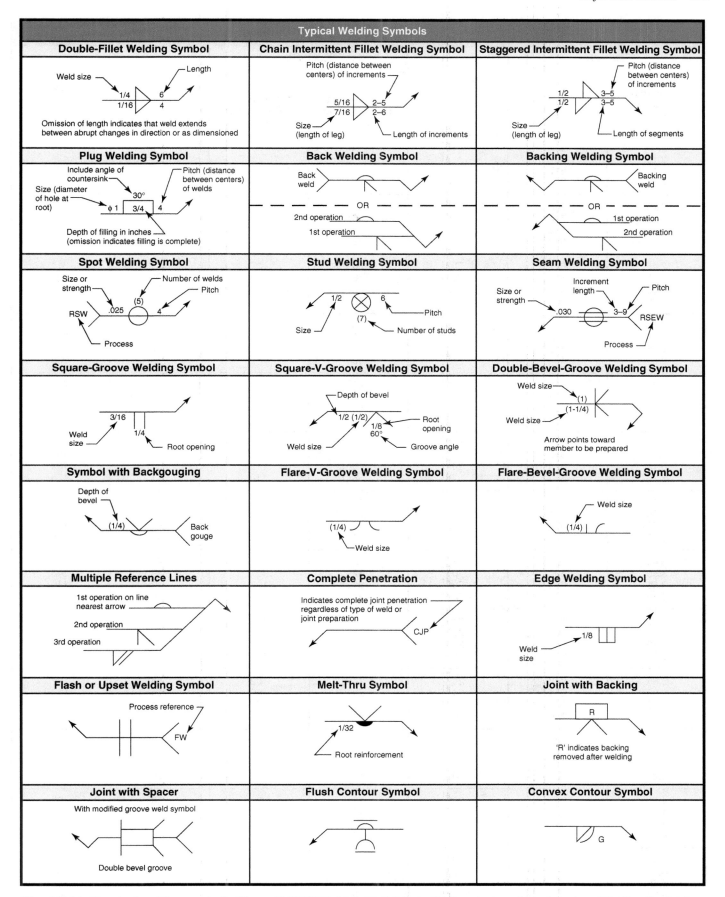

Chart R-16. Standard welding symbols. (From *ANSI/AWS A2.4-98*. Printed with Permission of the American Welding Society)

Hardness Conversion Table						
Brinell		Vickers or Firth Hardness No.	Rockwell		Scleroscope	Tensile Strength 1000 psi
Dia. in mm, 3000 kg. Load 10 mm Ball	Hardness No.		C 150 kg. Load 120° Diamond Cone	B 100 kg. Load 1/16 in. dia. Ball		
2.05	898					440
2.10	857					420
2.15	817					401
2.20	780	1150	70		106	384
2.25	745	1050	68		100	368
2.30	712	960	66		95	352
2.35	682	885	64		91	337
2.40	653	820	62		87	324
2.45	627	765	60		84	311
2.50	601	717	58		81	298
2.55	578	675	57		78	287
2.60	555	633	55	120	75	276
2.65	534	598	53	119	72	266
2.70	514	567	52	119	70	256
2.75	495	540	50	117	67	247
2.80	477	515	49	117	65	238
2.85	461	494	47	116	63	229
2.90	444	472	46	115	61	220
2.95	429	454	45	115	59	212
3.00	415	437	44	114	57	204
3.05	401	420	42	113	55	196
3.10	388	404	41	112	54	189
3.15	375	389	40	112	52	182
3.20	363	375	38	110	51	176
3.25	352	363	37	110	49	170
3.30	341	350	36	109	48	165
3.35	331	339	35	109	46	160
3.40	321	327	34	108	45	155
3.45	311	316	33	108	44	150
3.50	302	305	32	107	43	146
3.55	293	296	31	106	42	142
3.60	285	287	30	105	40	138
3.65	277	279	29	104	39	134
3.70	269	270	28	104	38	131
3.75	262	263	26	103	37	128
3.80	255	256	25	102	37	125
3.85	248	248	24	102	36	122
3.90	241	241	23	100	35	119
3.95	235	235	22	99	34	116
4.00	229	229	21	98	33	113
4.05	223	223	20	97	32	110
4.10	217	217	18	96	31	107
4.15	212	212	17	96	31	104
4.20	207	207	16	95	30	101
4.25	202	202	15	94	30	99
4.30	197	197	13	93	29	97
4.35	192	192	12	92	28	95
4.40	187	187	10	91	28	93
4.45	183	183	9	90	27	91
4.50	179	179	8	89	27	89
4.55	174	174	7	88	26	87
4.60	170	170	6	87	26	85

Chart R-17. Hardness table.

(Continued)

Hardness Conversion Table						
Brinell		**Vickers or Firth Hardness No.**	**Rockwell**		**Scleroscope**	**Tensile Strength 1000 psi**
Dia. in mm, 3000 kg. Load 10 mm Ball	**Hardness No.**		**C 150 kg. Load 120° Diamond Cone**	**B 100 kg. Load 1/16 in. dia. Ball**		
4.65	166	166	4	86	25	83
4.70	163	163	3	85	25	82
4.75	159	159	2	84	24	80
4.80	156	156	1	83	24	78
4.85	153	153		82	23	76
4.90	149	149		81	23	75
4.95	146	146		80	22	74
5.00	143	143		79	22	72
5.05	140	140		78	21	71
5.10	137	137		77	21	70
5.15	134	134		76	21	68
5.20	131	131		74	20	66
5.25	128	128		73	20	65
5.30	126	126		72		64
5.35	124	124		71		63
5.40	121	121		70		62
5.45	118	118		69		61
5.50	116	116		68		60
5.55	114	114		67		59
5.60	112	112		66		58
5.65	109	109		65		56
5.70	107	107		64		56
5.75	105	105		62		54
5.80	103	103		61		53
5.85	101	101		60		52
5.90	99	99		59		51
5.95	97	97		57		50
6.00	95	95		56		49

Chart R-17. *(Continued)*

Safe Limits for Welding Fumes							
Material	**Gases**		**Million Parts**	**Material**	**Gases**		**Million Parts**
	ppm	**mg/m³**	**per cu. ft.**		**ppm**	**mg/m³**	**per cu. ft.**
Acetylene	1000			Titanium oxide		15.0	
Beryllium		0.002		Zinc oxide fumes		5.0	
Cadmium oxide fumes		0.1		Silica, crystalline			2.5
				Silica, amorphous			20.0
Carbon dioxide	5000			Silicates:			
Copper fumes		0.1		Asbestos			5.0
Iron oxide fumes		10.0		Portland cement			50.0
Lead		0.2		Graphite			15.0
Manganese		5.0		Nuisance dust			50.0
Nitrogen dioxide	5.0						
Oil mist		5.0					

Chart R-18. Safe limits for welding fumes.

Elements	Symbol	Density (Specific Gravity)	Weight per cu. ft.	Specific Heat	Melting Point	
					°C	°F
Aluminum	Al	2.7	166.7	0.212	658.7	1217.7
Antimony	Sb	6.69	418.3	0.049	630	1166
Armco iron	. . .	7.9	490.0	0.115	1535	2795
Carbon	C	2.34	219.1	0.113	3600	6512
Chromium	Cr	6.92	431.9	0.104	1615	3034
Columbium	Cb	7.06	452.54	. . .	1700	3124
Copper	Cu	8.89	555.6	0.092	1083	1981.4
Gold	Au	19.33	1205.0	0.032	1063	1946
Hydrogen	H	0.070*	0.00533	. . .	−259	−434.2
Iridium	Ir	22.42	1400.0	0.032	2300	4172
Iron	Fe	7.865	490.9	0.115	1530	2786
Lead	Pb	11.37	708.5	0.030	327	621
Manganese	Mn	7.4	463.2	0.111	1260	2300
Mercury	Hg	13.55	848.84	0.033	−38.7	−37.6
Nickel	Ni	8.80	555.6	0.109	1452	2645.6
Nitrogen	N	0.97*	0.063	. . .	−210	−346
Oxygen	O	1.10*	0.0866	. . .	−218	−360
Phosphorus	P	1.83	146.1	0.19	44	111.2
Platinum	Pt	21.45	1336.0	0.032	1755	3191
Potassium	K	0.87	54.3	0.170	62.3	144.1
Silicon	Si	2.49	131.1	0.175	1420	2588
Silver	Ag	10.5	655.5	0.055	960.5	1761
Sodium	Na	0.971	60.6	0.253	97.5	207.5
Sulfur	S	1.95	128.0	0.173	119.2	246
Tin	Sn	7.30	455.7	0.054	231.9	449.5
Titanium	Ti	5.3	218.5	0.010	1795	3263
Tungsten	W	17.5	1186.0	0.034	3000	5432
Uranium	U	18.7	1167.0	0.028		
Vanadium	V	6.0	343.3	0.115	1720	3128
Zinc	Zn	7.19	443.2	0.093	419	786.2
Bronze (90% Cu 10% Sn)	. . .	8.78	548.0	. . .	850–1000	1562–1832
Brass (90% Cu 10% Zn)	. . .	8.60	540.0	. . .	1020–1030	1868–1886
Brass (70% Cu 30% Zn)	. . .	8.44	527.0	. . .	900–940	1652–1724
Cast pig iron	. . .	7.1	443.2	. . .	1100–1250	2012–2282
Open-hearth steel	. . .	7.8	486.9	. . .	1350–1530	2462–2786
Wrought-iron bars	. . .	7.8	486.9	. . .	1530	2786

*Density compared with air.

Chart R-19. Properties of elements and metal compositions.

General Metric/U.S. Conventional Conversions			
Property	**To Convert from**	**To**	**Multiply by**
Acceleration (angular)	revolution per minute squared	rad/s^2	$1.745\ 329 \times 10^{-3}$
Acceleration (linear)	in/min^2 ft/min^2 in/min^2 ft/min^2 ft/s^2	m/s^2 m/s^2 mm/s^{2}mm/s^2 m/s^2 m/s^2	$7.055\ 556 \times 10^{-6}$ $8.466\ 667 \times 10^{-5}$ $7.055\ 556 \times 10^{-3}$ $8.466\ 667 \times 10^{-2}$ $3.048\ 000 \times 10^{-1}$
Angle, plane	deg minute second	rad rad rad	$1.745\ 329 \times 10^{-2}$ $2.908\ 882 \times 10^{-4}$ $4.848\ 137 \times 10^{-6}$
Area	in^2 ft^2 yd^2 in^2 ft^2 acre (U.S. Survey)	m^2 m^2 m^2 mm^2 mm^2 m^2	$6.451\ 600 \times 10^{-4}$ $9.290\ 304 \times 10^{-2}$ $8.361\ 274 \times 10^{-1}$ $6.451\ 600 \times 10^{2}$ $9.290\ 304 \times 10^{4}$ $4.046\ 873 \times 10^{3}$
Density	pound mass per cubic inch pound mass per cubic foot	kg/m^3 kg/m^3	$2.767\ 990 \times 10^{4}$ $1.601\ 846 \times 10$
Energy, work, heat, and impact energy	foot pound force foot poundal Btu* calorie* watt hour	J J J J J	$1.355\ 818$ $4.214\ 011 \times 10^{-2}$ $1.054\ 350 \times 10^{3}$ $4.184\ 000$ $3.600\ 000 \times 10^{3}$
Force	kilogram-force pound-force	N N	$9.806\ 650$ $4.448\ 222$
Impact strength	(see Energy)		
Length	in ft yd rod (U.S. Survey) mile (U.S. Survey)	m m m m km	$2.540\ 000 \times 10^{-2}$ $3.048\ 000 \times 10^{-1}$ $9.144\ 000 \times 10^{-1}$ $5.029\ 210$ $1.609\ 347$
Mass	pound mass (avdp) metric ton ton (short, 2000 lb/m) slug	kg kg kg kg	$4.535\ 924 \times 10^{-1}$ $1.000\ 000 \times 10^{3}$ $9.071\ 847 \times 10^{2}$ $1.459\ 390 \times 10$
Power	horsepower (550 ft lbs/s) horsepower (electric) Btu/min* calorie per minute* foot pound-force per minute	W W W W W	$7.456\ 999 \times 10^{2}$ $7.460\ 000 \times 10^{2}$ $1.757\ 250 \times 10$ $6.973\ 333 \times 10^{-2}$ $2.259\ 697 \times 10^{-2}$
Pressure	pound force per square inch bar atmosphere kip/in^2	kPa kPa kPa kPa	$6.894\ 757$ $1.000\ 000 \times 10^{2}$ $1.013\ 250 \times 10^{2}$ $6.894\ 757 \times 10^{3}$
Temperature	degree Celsius, $t°_C$ degree Fahrenheit, $t°_F$ degree Rankine, $t°_R$ degree Fahrenheit, $t°_F$ kelvin, t_K	K K °R °C °C	$t_K = t°_C + 273.15$ $t_K = (t°_F + 459.67)/1.8$ $t_K = t°_R/1.8$ $t°_C = (t°_F - 32)/1.8$ $t°_C = t_K - 273.15$
Tensile strength (stress)	ksi	MPa	$6.894\ 757$
Torque	inch pound force foot pound force	N·m N·m	$1.129\ 848 \times 10^{-1}$ $1.355\ 818$

*Thermochemical

Chart R-20. Metric to U.S. Conventional conversions.

(Continued)

General Metric/U.S. Conventional Conversions			
Property	**To Convert From**	**To**	**Multiply By**
Velocity (angular)	revolution per minute degree per minute revolution per minute	rad/s rad/s deg/min	$1.047\ 198 \times 10^{-1}$ $2.908\ 882 \times 10^{-4}$ $3.600\ 000 \times 10^{2}$
Velocity (linear)	in/min ft/min in/min ft/min mile/hour	m/s m/s mm/s mm/s km/h	$4.233\ 333 \times 10^{-4}$ $5.080\ 000 \times 10^{-3}$ $4.233\ 333 \times 10^{-1}$ $5.080\ 000$ $1.609\ 344$
Volume	in³ ft³ yd³ in³ ft³ in³ ft³ gallon	m³ m³ m³ mm³ mm³ L L L	$1.638\ 706 \times 10^{-5}$ $2.831\ 685 \times 10^{-2}$ $7.645\ 549 \times 10^{-1}$ $1.638\ 706 \times 10^{4}$ $2.831\ 685 \times 10^{7}$ $1.638\ 706 \times 10^{-2}$ $2.831\ 685 \times 10$ $3.785\ 412$

Chart R-20. *(Continued)*

Metric Units for Welding		
Property	**Unit**	**Symbol**
Area dimensions	Square millimeter	mm²
Current density	Ampere per square millimeter	A/mm²
Deposition rate	Kilogram per hour	kg/h
Electrical resistivity	Ohm meter	$\Omega \cdot m$
Electrode force (upset, squeeze, hold)	Newton	N
Flow rate (gas and liquid)	Liter per minute	L/min
Fracture toughness	Meganewton meter$^{-3/2}$	MN·m$^{-3/2}$
Impact strength	Joule	J = N·m
Linear dimensions	Millimeter	mm
Power density	Watt per square meter	W/m²
Pressure (gas and liquid)	Kilopascal	kPa = 1000 N/m²
Tensile strength	Megapascal	MPa = 1000 000 N/m²
Thermal conductivity	Watt per meter Kelvin	W/(m·K)
Travel speed	Millimeter per second	mm/s
Volume dimensions	Cubic millimeter	mm³
Electrode feed rate	Millimeter	mm/s

Chart R-21. Metric units for welding.

Converting Measurements for Common Welding Properties			
Property	**To convert from**	**To**	**Multiply by**
Area dimensions (mm^2)	in^2 mm^2	mm^2 in^2	$6.451\ 600 \times 10^2$ $1.550\ 003 \times 10^{-3}$
Current density (A/mm^2)	A/in^2 a/mm^2	A/mm^2 A/in^2	$1.550\ 003 \times 10^{-3}$ $6.451\ 600 \times 10^2$
Deposition rate** (kg/h)	lb/h kg/h	kg/h lb/h	0.045** 2.2**
Electrical resistivity (Ω·m)	Ω·cm Ω·m	Ω·m Ω·cm	$1.000\ 000 \times 10^{-2}$ $1.000\ 000 \times 10^2$
Electrode force (N)	pound-force kilogram-force N	N N ibf	4.448 222 9.806 650 $2.248\ 089 \times 10^{-1}$
Flow rate (L/min)	ft^3/h gallon per hour gallon per minute cm^3/min L/min cm^3/min	L/min L/min L/min L/min ft^3/min ft^3/min	$4.719\ 475 \times 10^{-1}$ $6.309\ 020 \times 10^{-2}$ 3.785 412 $1.000\ 000 \times 10^{-3}$ 2.118 880 $2.118\ 880 \times 10^{-3}$
Fracture toughness (MN·m$^{-3/2}$)	ksi·in$^{1/2}$ MN·m$^{-3/2}$	MN·m$^{-3/2}$ ksi·in$^{1/2}$	1.098 855 0.910 038
Heat input (J/m)	J/in J/m	J/m J/in	$3.937\ 008 \times 10$ $2.540\ 000 \times 10^{-2}$
Impact energy	foot pound force	J	1.355 818
Linear measurements (mm)	in ft mm mm	mm mm in ft	$2.540\ 000 \times 10$ $3.048\ 000 \times 10^2$ $3.937\ 008 \times 10^{-2}$ $3.280\ 840 \times 10^{-3}$
Power density (W/m^2)	W/in^2 W/m^2	W/m^2 W/in^2	$1.550\ 003 \times 10^3$ $6.451\ 600 \times 10^{-4}$
Pressure (gas and liquid) (kPa)	psi lb/ft^2 N/mm^2 kPa kPa kPa torr (mm Hg at 0°C) micron (μm Hg at 0°C) kPa kPa	Pa Pa Pa psi il/ft^2 N/mm^2 kPa kPa torr micron	$6.894\ 757 \times 10^3$ $4.788\ 026 \times 10$ $1.000\ 000 \times 10^6$ $1.450\ 377 \times 10^{-1}$ $2.088\ 543 \times 10$ $1.000\ 000 \times 10^{-3}$ $1.333\ 22 \times 10^{-1}$ $1.333\ 22 \times 10^{-4}$ $7.500\ 64 \times 10$ $7.500\ 64 \times 10^3$
Tensile strength (MPa)	psi lb/ft^2 N/mm^2 MPa MPa MPa	kPa kPa MPa psi lb/ft^2 N/mm^2	6.894 757 $4.788\ 026 \times 10^{-2}$ 1.000 000 $1.450\ 377 \times 10^2$ $2.088\ 543 \times 10^4$ 1.000 000
Thermal conductivity (W[m·K])	cal/(cmÅs·°c)	W/(m·K)	$4.184\ 000 \times 10^2$
Travel speed, electrode feed speed (mm/s)	in/min mm/s	mm/s in/min	$4.233\ 333 \times 10^{-2}$ 2.362 205

*Preferred units are given in parentheses.
**Approximate conversion.

Chart R-22. Converting measurements for common welding properties.

Glossary of Welding Terms

A

Abrasive pads: Commercial cleaning pads used to remove oxide film from the weld joint area before welding.

Acetone: Colorless, volatile, water-soluble, flammable liquid used to remove grease and oils from the weld joint before welding.

Acid pickle: Chemical bath of various acids used to remove heavy oxide scale and foreign materials from the weld area.

Afterflow time: Period of time after welding in which shielding gas flows through the welding gun nozzle to shield the electrode from contamination.

Age: Metallurgical term to describe a natural hardening sequence that changes the mechanical properties of a material.

Air-carbon arc cutting (AAC): Arc cutting process in which the metals to be cut are melted by the heat of the carbon arc. Molten metal is removed by a blast of compressed air.

Air-cooled gun: Used with a self-shielded electrode. Has a small contact tip insulator on the end that does not need to be removed for welding.

Alkaline cleaner: Mixture of sodium hydroxide and water used for removing oil and oxides from metal. Commonly used on aluminum and magnesium for preweld cleaning.

Alloy: A material formed by the combination of two or more metallic elements.

Alternating current (ac): Electrical current is the flow of electrons through a conductor. In alternating current the electrons flow in one direction, stop, and reverse the flow (alternate directions) for each complete cycle. In the United States, alternating current flows at 60 cycles per second. Therefore, the current stops and starts 120 times per second.

Alternator: Refers to an alternating current generator.

American Iron and Steel Institute (AISI): Industry association of iron and steel producers. It provides statistics on steel production and use, and publishes steel products manuals.

American Society for Testing Materials (ASTM): Organization that writes specifications for iron and steel products for industry.

American Welding Society (AWS): Nonprofit technical society organized and founded for the purpose of advancing the art and science of welding. The AWS publishes codes and standards concerning all phases of welding and *The Welding Journal.*

Ammeter: Instrument for measuring electrical current in amperes.

Amperage: Strength of an electrical current measured in amperes.

Anneal: Removal of internal stresses in metal by heating and slow cooling.

Anode: Positive terminal or pole of a circuit.

Arc blow: Deflection of the intended arc pattern by magnetic fields.

Arc gap: Welding voltage. *See* arc length.

Arc length: Distance from the end of the electrode to the weld pool. Also called *arc gap.*

Arc voltage: Measurement of electrical potential (pressure or voltage) across the arc gap between the end of the electrode and the workpiece during the welding process.

Arc welding gun: A device used in semiautomatic, machine, and automatic arc welding to transfer current, guide the consumable electrode, and direct the shielding gas.

Argon: Inert gas used in FCAW to shield the electrode and weld metal.

Armature: Part of an electric machine that includes the main current-carrying windings.

Artificially aged: A metallurgical term that describes a heating operation in the hardening sequence that changes the mechanical properties of metal.

ASME Boiler and Pressure Vessel Code: Various sections cover the design, construction, and inspection of boilers and pressure vessels. Section IX of the code covers material specifications, nondestructive examinations, and welding qualifications.

Asphyxiation: Loss of consciousness due to a lack of oxygen.

Atmosphere: Envelope in which the welding arc is enclosed. Also called inert atmosphere.

Austenitic stainless steels: Iron alloys that contain at least 11% chromium with varying amounts of nickel. The grain structure is nonmagnetic (austenitic).

Automatic operation: Welding that is mechanically moved while controls govern the speed and direction of the weld process.

Automatic welding: Welding with equipment that performs the welding operation without adjustment of the controls by a welding operator. The equipment may or may not load and unload the workpiece.

Auxiliary shielding gas: Protects the arc stream and molten pool from atmospheric oxygen and nitrogen.

B

Backflow check valve: Prevents backflow and improper gas mixing.

Backhand welding: A welding technique in which the welding electrode is directed opposite the progression of welding.

Backing: A material or device placed against the back side of a weld joint. The material may be partially fused or remain unfused during welding, and it may be either a metal or nonmetal.

Backing bar: Tool or fixture attached to the root of the weld joint. It may or may not control the shape of the penetrating metal.

Backstep sequence: A longitudinal sequence in which weld passes are made in the direction opposite the progression of welding.

Base material: The material to be welded.

Bevel: Angular type of edge preparation.

Bore: Inside diameter of a hole, tube, or hollow object.

Brass: Various metal alloys consisting mainly of copper and zinc.

Bright anneal: Process of annealing (softening) metal, usually carried out in a controlled furnace atmosphere so surface oxidization is reduced to a minimum. The surface remains relatively bright.

Bright metal: Material preparation in which the surface is ground or machined to a bright surface to remove scale or oxides.

Brittle weld: Hard weld with little or no ductility.

Bronze: Various metal alloys consisting mainly of copper and tin.

Buildup: A surface variation in which surfacing metal is deposited to achieve the required dimensions.

Burn through: Weld that has melted through, resulting in a hole and excessive penetration.

Buttering: Depositing surfacing metal on one or more surfaces to provide metallurgically compatible weld metal for completion of the weld.

Butt joint: A joint between two members aligned approximately in the same plane.

C

Cadmium: White, ductile, metallic element used for plating material to prevent corrosion. The fumes generated by heating cadminum plating are toxic.

Calibration: A means to determine, check, or rectify potentiometer output.

Carbide: Compound of carbon with one or more metal elements.

Carbide precipitation: Movement of chromium from the grains into the grain boundaries in chrome-nickel stainless steel.

Carbon dioxide: A compound of carbon and oxygen used as a primary shielding gas or part of a mixed shielding gas combination. Symbol is CO_2.

Cast: The amount of curvature in a coil of electrode wire.

Cathode: Negative terminal or pole of a circuit.

Certified chemical analysis: Report of chemical analysis on a particular heat, lot, or section of material.

Charpy impact test: Test used to determine resistance to failure caused at a notch at low temperatures.

Chemical composition: Composition of material in chemical percentages.

Cladding: Depositing surfacing material to improve corrosion- or heat-resistance.

Cold lap: Area of a weld that has not fused with the base material.

Cold work: To increase the tensile properties of a material by working the material without heat.

Complete joint penetration: Penetration by weld metal for the full thickness of the base metal in a joint with a groove weld.

Concave weld crown: Weld crown that is curved inward.

Conduit: Metallic sheath or tube through which the electrode is moved from the wire feeder to the weld area.

Constant current: An arc welding power source with a volt-ampere relationship, yielding a small welding current change from a large arc voltage change.

Constant voltage: A welding power source with a volt-ampere relationship yielding a large welding current change from a small arc voltage change.

Contact tip: Component of a gun that conducts electrical current from the power supply to the consumable welding electrode.

Contactor: Electric switch in a power supply. Activating the contactor transfers the welding power from the welding machine to the welding gun.

Contamination: Indicates a dirty part, impure argon, or improper gas shielding.

Contour: Shape of the weld bead or pass.

Convex weld crown: Weld crown that is curved outward.

Copper: Malleable, ductile, metallic element.

Corrective action: Action to be taken to prevent weld discrepancies.

Corrosion: Eating away of material by a corrosive medium.

Corrosion-resistant: Properties of a metal that resist chemical or electrochemical interaction with the surroundings, preventing the metal from deterioration.

Crack: A fracture-type discontinuity characterized by a sharp tip and high ratio of length and width to opening displacement.

Cracking: Blowing dirt out of a high-pressure gas cylinder valve.

Crater: Depression at the end of a weld that has insufficient cross section.

Crater cracks: Cracking that occurs in the crater.

Cryogenic temperature: Very cold temperatures. Usually considered to be the temperature of liquefied gases.

Cubic feet per hour (cfh): Measurement of the amount of gas flow used in FCAW operations.

D

Defects: Extensive flaws in a weldment that could cause weld failure during service.

Deionized water: Water that has been specially treated for removal of chemicals.

Demurrage: Monetary charge applied to the user of gas cylinders beyond the agreed rental period.

Department of Transportation (DOT): Government organization responsible for establishing and maintaining rules and precautions for the safe handling of fuels and gases used in welding.

Deposition rate: The weight of material deposited in a unit of time.

Depth of fusion: Distance that fusion extends into the base metal or previous pass from the surface melted during welding.

Destructive testing: Series of tests by destruction to determine the quality of a weld.

Dewar cylinder: Specially constructed tank or flask, similar to a vacuum bottle, for the storage of liquefied gases.

Dies: Tools used to form wire or metal to a specified shape or dimension.

Digital: Use of numerical digits to establish operational limits.

Direct current (dc): Flow of current (electrons) in one direction, either to the workpiece or to the electrode.

Direct current electrode negative (DCEN): Direct current flowing from the electrode to the work. Also called direct current straight polarity (DCSP).

Direct current electrode positive (DCEP): Direct current flowing from the work to the electrode. Also called direct current reverse polarity (DCRP).

Direct current reverse polarity (DCRP): See direct current electrode positive.

Direct current straight polarity (DCSP): See direct current electrode negative.

Discontinuities: Welding flaws.

Dissipate heat: Remove heat from the weld zone by the use of jigs or fixtures.

Drag angle: Travel angle when the electrode is pointed in a direction opposite the progression of welding (backhand or pull welding).

Drive rollers: Specially designed rollers for various types and sizes of electrodes to be fed through a mechanized wire feeder.

Dross: Oxidized metal or impurities within the metal.

Ductility: Property of material allowing it to deform or exhibit plasticity without breaking under tension.

Duty cycle: The percentage of time during an arbitrary test period that a power source or its accessories can be operated at rated output without overheating.

E

Effective throat: The minimum distance minus any convexity between the weld root and the face of a fillet weld.

Electrode stickout (ESO): The distance from the end of the contact tip to the end of the electrode.

Electronic potentiometers: Electrical devices using electronic circuits to control and regulate electrical current flow.

Electrons: Negatively charged particles.

Elongation: Permanent elastic extension of metal during tensile testing. The amount of extension is usually indicated by percentages of original gauge length.

Extra-low interstitial: Very small amount of carbon, oxygen, hydrogen, and nitrogen.

F

Feathered tack weld: Tack weld tapered on both ends to assist in obtaining the proper penetration during a groove joint root pass weld.

Ferrite test: A test of austenitic stainless steel weld deposit to determine the amount of ferrite.

Ferritic stainless steel: Group of iron-chromium and carbon alloys that are nonhardenable by heat treatment.

Ferromagnetic: Material that can possess magnetization in the absence of an external magnetic field.

Ferrous metals: Group of metals containing substantial amounts of iron.

Fillet weld: Triangular cross-section weld joining two surfaces approximately at right angles to each other in a lap joint, T-joint, or corner joint.

Fillet weld leg: Leg lengths of the largest isosceles right triangle that can be inscribed within a fillet weld cross section.

Fillet weld throat: Distance measured from the bottom corner of the weld joint to the crown surface.

Flowmeter: Mechanical device used to regulate the amount of gas flow to the weld area.

Flux cored arc welding (FCAW): An arc welding process that produces coalescence of metals by heating them with an arc between a continuous filler metal electrode and the work. Shielding is provided by a flux contained within the electrode. Additional shielding may or may not be obtained from an externally supplied gas or gas mixture.

Flux cored electrode: A composite filler metal electrode consisting of a metal tube or other hollow configuration and containing ingredients to provide such functions as shielding atmosphere, deoxidization, arc stabilization, and slag formation. Minor amounts of alloying material may be included in the core. External shielding may or may not be used.

Forehand welding: A welding technique in which the electrode is pointed toward the progression of welding. Also called *push welding*.

Fuses: Electrical circuit device designed to fail at a predetermined current level to provide protection from overcurrent.

Fusion: Melting together of filler metal and base metal, or of base metal only.

Fusion welding: A process that uses fusion to complete the weld.

G

Gas-cooled gun: Used with a gas-shielded electrode. Has an adapter on the end for the gas nozzle.

Gas envelope: Shape and pattern of shielding gas over the weld area.

Gas nozzle: An assembly made from glass, metal, or ceramics of various designs. Attaches to the gun body to direct shielding gas flow (envelope) over the weld area.

Gas-shielded arc process: Welding method in which a gas or combination of gases shields and protects the molten stream and weld metal from the atmosphere.

Gas-shielded flux cored arc welding (FCAW-g): A flux cored arc welding process variation in which shielding gas is supplied through the gas nozzle, in addition to the shielding that is obtained from the flux within the electrode.

Globular arc mode: Method of weld metal deposition as large globules of metal that detach from the electrode in an indefinite pattern.

Globular transfer: The transfer of molten metal in large drops from a consumable electrode across the arc.

Grids: Welder control station with adjustable potentiometers or switches on a multioperator welding system.

Grit blasting: Process for cleaning or finishing metal using an air blast that blows particles of an abrasive (small pieces of steel, sand, or steel balls) against the workpiece.

Groove angle: The total included angle of the groove between the workpieces.

Gun: Trigger or switch-operated mechanism held by an operator or mounted in machine that transfers electrical current to the electrode. It may contain a gas nozzle to direct shielding gas around the molten pool.

Gun cable: Covered bundle of hoses and conductors used to carry electrical current, gas, welding electrode, cooling water, electrical circuit wire (from the start switch), and vacuum hoses (for smoke removal) to the gun.

H

Hardfacing: A surfacing variation in which hard material is deposited on the surface of softer material for protection from abrasion or wear.

Hard surfacing: Hard material applied to the surface of softer material for protection from abrasion or wear.

Hard tooling: Specially designed tooling or fixturing to hold the parts of a weldment during a welding operation.

Heat-affected zone: Portion of the base metal that has not been melted, but whose mechanical properties or microstructure have been altered by the heat of welding.

Heat-resistant: Materials that are resistant to oxidation at high temperatures.

Heat sinks: Tooling applied adjacently to the weld zone to absorb heat and prevent heat flow into the parent material.

Hot shortness: A weakness exhibited by aluminum and some other metals (low strength level) when hot.

Hydraulic systems: Pressure systems that use fluids, such as oil, to move or hold weld parts or tooling.

Hydromount: A portable welding machine cart that has a supply of cooling water for the welding gun.

Hydrostatic pressure: Pressure obtained by water.

I

Icicles: Intermittent sections of weld drop-through extending below the normal contour of a full-penetration groove weld.

Impurity level: Level of impurities established in metal and gases to reduce contamination of the weld.

Inconel: Nickel alloys that contain substantial percentages of chromium and iron.

Inert gas: Gas that does not normally combine chemically with base metal or filler material.

Ingot: Large block of metal usually cast in a metal mold that forms the basic material for further processing.

Intergranular corrosion: Occurs when welded austenitic stainless steels have been sensitized during welding, then subjected to some type of acid solution to induce corrosion.

Interpass temperature: The minimum/maximum temperature of the weld metal before the next pass is made in a multiple-pass weld.

Ionization potential: Energy required to remove an electron from an atom, making it an ion. Shielding gases have different ionization potentials.

Izod test: Test used to determine notch impact values. Often used in conjunction with Charpy impact test.

J

Joint: Area where two or more pieces are brought together in an assembly.

Joint design: The shape, dimensions, and configuration of the joint.

Joint efficiency: The ratio of strength of a joint to the strength of the base metal, expressed in a percentage.

K

Kinetic energy: Energy of a body or system with respect to the motion of the body or the particles in the system.

L

Lack of fusion: Fusion that is less than complete.

Lack of penetration: Joint penetration that is less than specified.

Linear porosity: Cavity-type discontinuities formed by gas entrapment along a line during solidification of liquid melt.

Liquefied gas: Gas that has been changed into a liquid for ease of storage and handling.

Liquid penetrant inspection: Visual surface inspection using dye and developer. May also use fluorescent dye and black light for observing discontinuities in a weld.

Liters per minute: Metric measurement of the amount of gas flow used in the welding operation.

Long stickout: Visible stickout distance used when making high-deposition welds in flat or horizontal positions in which the electrode is preheated for more rapid melting.

M

Machine duty cycle: Period of time established by the machine manufacturer for operation within the machine's design specification.

Macroetch test: A test in which the specimen is prepared with a fine finish, etched with an acid to define the grain structure, and examined under low magnification.

Macrotest: Visual test of a weld or parent metal structure; the cross section using low magnification or the naked eye.

Magne gauge test: Inspection tool designed to test stainless steel welds for ferrite content.

Magnetic field: Area where the magnetic lines of force have been established.

Magnetic particle inspection: Nondestructive inspection test that uses iron particles to outline discontinuities on a magnetized weld or a parent metal.

Manifold: Pipe or cylinder with several inlet and outlet fittings that allows several cylinders to be fitted together to supply multiple welding stations.

Manipulator: Positioning tool used to locate welding equipment at the weld area for longseam or circumferential welding.

Manual welding: Welding operation performed manually.

Martensite: Structure obtained when steel is heated and cooled to achieve its maximum hardness.

Mechanical properties: Includes such areas as tensile strength, ductility, brittleness, elasticity, hardness, toughness, and malleability.

Metal cored electrode: A composite filler metal electrode consisting of a metal tube or other hollow configuration containing alloying materials. Minor amounts of ingredients providing such functions as arc stabilization and fluxing of oxides may be included. External shielding gas may or may not be used.

Microcracks: Very small cracks within metal that are located and seen only with the aid of very high magnification.

Microswitch: Used in foot or hand controls for initiating welding cycles; the switch operates by depressing a lever a few thousandths of an inch.

Mismatch: Junction point of a butt joint where the top or bottom edges are not even.

Mode: Setup of a welding machine for manual, remote, or automatic operation.

Module: Unit that contains all the required circuits for a specified sequence.

Motor generator: A welding power supply that uses an electrical motor to drive a direct current welding generator.

Multioperator systems: Master power supply that provides power to a series of welder-controlled grids or control panels.

N

National Electrical Manufacturers Association (NEMA): Industrial association of manufacturers of electrical machinery. Publishes standards and industry statistics, including those for welding.

Neoprene: Oil-resistant synthetic rubber.

Nick break test: Destructive test of the weld to determine internal quality.

Nitrogen: Colorless, odorless, gaseous element.

Nondestructive testing: Testing that does not require destroying the part to determine quality.

Nonferrous: Any metal that does not contain iron.

Nonprequalified joints: Joints not made with materials that have been approved by a governing welding code.

Nontoxic: Not harmful or poisonous.

Notch sensitivity: Resistance of metal to notch failure when subjected to rapid loading or stress.

Notch toughness: Ability of metal to resist failure (cracking) at a notch during loading (stress).

Nozzle: Device that attaches to the end of the welding gun and directs the flow of shielding gas.

Numerical control: Numerical program established on tape or disk that controls the parameters and sequences of the welding operation.

Nylon: Thermoplastic polyamide that can be molded into welding gun parts of extreme toughness, strength, and elasticity.

O

Opaque: Material that does not allow light to pass through; not pervious to light.

Open arc process: Welding method in which all the fluxing ingredients are included in the core material to protect the molten stream and weld metal from the atmosphere.

Open circuit voltage: The voltage between the output terminals of the power source when no current is flowing to the gun.

Oscillate: Moving welding gun back and forth across weld joint.

Oscillated beads: *See* wash beads.

Out-of-position welding: Welding performed in a nonstandard position such as flat, vertical, and overhead.

Output rating: Output limits of a power supply with regard to current, open circuit voltage, ranges, power factor, and duty cycle.

Ovality: Amount of out-of-round condition when referring to pipe, tubing, or round objects.

Overlap: Protrusion of weld metal beyond the toe, face, or root of the weld.

Overlay: Weld placed over the top of another metal for dimensional requirements or to add physical or mechanical properties.

Oxidation: Process of reaction with an oxidizing agent.

Oxide film: Film formed on base material as a result of exposure to oxidizing agents, atmosphere, chemicals, or heat.

Oxygen analyzer: Instrument used to determine the amount of oxygen within an area, measured in parts per million (ppm).

P

Parameters: The range of operating values specified on the Welder Procedure Specification. These include amperage, voltage, travel speed, shielding gas, and flow rate (if required).

Parts per million (ppm): Numerical method of identifying the purity or contamination of a gas.

Passes: Single-weld beads.

Penetrameters: Metal shims with specified hole sizes used during the exposure of radiographic film to verify sensitivity and proper exposure.

Penetration: Minimum depth a groove or flange weld extends from its face into the joint, exclusive of reinforcement.

Perimeter: Outer edge of a circumferential part.

Physical properties: Properties of metals relating to electrical, thermal conductivity, and expansion rates.

Plasma: Gas that has been heated to a partially ionized condition, enabling it to conduct an electrical current.

Plasma-arc cutting: Arc-cutting process that severs or cuts metal by melting a localized area with a constricted arc. A process to remove the molten metal with a high velocity jet of hot, ionized gas issued from the gun orifice.

Polarity: Direction of current flow. Current flow from the electrode to the workpiece is DCEN or DCSP. Current flow from the workpiece to the electrode is DCEP or DCRP.

Pore: A single hole of indefinite size and shape located in the completed weld pass or layer of weld metal. The pore is usually caused by entrapped gas that formed during the melting of the material and was trapped in the cooling metal.

Porosity: Series of pores within a weld.

Positioner: Mechanical device for holding a workpiece for welding in the desired position. It may rotate the weldment with a controlled speed for circular welds.

Postflow: Period of time at the end of the weld cycle where shielding gas flows around the electrode during cooling to prevent contamination.

Postheating: Applying heat at the end of the weld cycle to slow down the cooling rate, prevent cracking, and relieve stresses.

Pounds per square inch (psi): Measurement of pressure.

Power supply: Machine designed to produce the type and amount of welding current necessary to melt the base metal and welding electrode.

Precipitation: Movement of elements within metal from the grain to the grain boundary.

Preheating: Applying heat to the weldment before starting the welding operation. The actual heating temperature will vary depending on the type, thickness, and condition of the material to be welded.

Preheat temperature: The temperature of the base metal in the volume surrounding the point of welding immediately before welding is started.

Primary voltage: Incoming voltage of alternating current supplied by a utility company.

Procedure qualification: Demonstration that welds made by a specific procedure can meet prescribed standards.

Procedure Qualification Record (PQR): Document providing the actual welding parameters and variables used to produce an acceptable test weld, and the results of tests conducted on the weld to qualify a welding procedure specification.

Programmer: Electronic or mechanical sequencer used to start and finish portions or all of the sequences required to complete a weld.

Pull welding: Method in which the welding gun is pointed back toward the molten pool. Also called *backhand welding.*

Pull-type feeder: Type of electrode feeder, usually located in the welding gun, that pulls the electrode through the cable to the contact tip.

Purging: Operation to remove air (atmosphere) from the welding area and replace it with an inert atmosphere.

Purging dams: Used in conjunction with the purging of pipe to reduce the area to be purged.

Push angle: The travel angle when the electrode is pointed in the direction of weld progression.

Push welding: Method in which the welding gun is pointed away from the molten pool. Also called *forehand welding.*

Push-type feeder: Type of electrode feeder in which the electrode is pushed from the feeder through the cable to the gun.

Push-pull-type feeder: Type of electrode feeder in which the electrode is pushed through the cable to a separate feeder mounted in the gun handle. Used for driving electrodes long distances at a constant rate of speed.

Q

Qualified procedure: Welding sequence that has met all the testing requirements of the fabrication specification.

Qualified welder: Person who has demonstrated the ability to produce a weld within the requirements of the fabrication specification.

Qualified welding operator: Person who has demonstrated the ability to operate a welding machine with a qualified procedure to produce a weld within the requirements of the fabrication specification.

R

Radiation: Energy that a welding arc radiates (sends out) in the form of intense ultraviolet light. Radiation can cause burns if the body is not adequately protected.

Radiographic inspection: Nondestructive testing method that uses X-rays and gamma rays to determine the interior quality of a weld.

Rated load: Load (welding current) that may be obtained from a power supply within limits established by the machine duty cycle.

Reactor core: Controls the rate of change of current in a circuit. Consists of an iron base with a wire wound around it.

Reagent: A high-purity substance, chemical, or solution used to test other substances, chemicals, or solutions for reactions.

Recrystallized: New grain structure obtained by heating a cold-worked metal.

Red-heat temperature: Temperature point in some metals where the material becomes brittle.

Reference arc voltage: Actual welding arc voltage established on the automatic welding control circuit. By reference to this voltage, the arc voltage-sensing control circuit maintains the desired arc voltage.

Reference line: The flat line on which a symbol is placed in a welding drawing.

Regulator: Valve used to reduce and control gas pressure from the cylinder to the gun. It keeps pressure constant.

Regulator/flowmeter: Mechanical device commonly used for high-pressure cylinders to control the flow of gas.

Reverse polarity: Electron flow is from workpiece to electrode. *See* direct current electrode positive.

Rheostat: Adjustable resistor whose resistance may be changed without opening the circuit in which it is connected.

Rimmed steel: Steel not completely deoxidized during manufacture. The outer rim of the ingot contains high-quality steel used to make steel welding electrodes.

Root opening: A separation at the joint root between workpieces.

Root penetration: Penetration that extends beyond the root of the welding joint.

Rotary selector: Mechanical method used by some power supply manufacturers to establish welding current output control.

Routing: Removal of metal from a weld using a routing tool. The tool may have various shapes and forms. Cutters may be tool steel or tungsten carbide.

Run-on/run-off tabs: Metal pieces used on longseam welds to obtain penetration at the start of the weld and to prevent craters at the end of the weld.

Rust inhibitor: Chemical added to water to inhibit (reduce) formation of rust particles.

S

Seamer: Machine designed to hold material for making longseam welds.

Seamwelder: Seamer machine with attached welding equipment for making longseam welds.

Self-shielded flux cored arc welding (FCAW-ss): A flux cored arc welding process variation in which shielding is obtained exclusively from the flux within the electrode.

Semiautomatic operation: Welding that is manually done by the welder controlling the gun position and travel speed while the special equipment feeds the electrode and/or shielding gas as required.

Semiautomatic welding: Welding operation where the welding operator controls the sequences and adjusts various parameters as required.

Semikilled steel: Steel that has been partially deoxidized during solidification in the ingot mold.

Sensitization range: Temperature range in stainless steels where carbide precipitation can take place.

Severn gauge: Inspection tool designed to test stainless steel welds for ferrite content.

Shielding gas: Gas used to shield molten metal from atmospheric contamination.

Shop-aid tooling: Clamps, screws, and metal forms used to hold parts in alignment for welding.

Short circuit: The arc is started by touching the electrode to the grounded weldment. It is often used when no high-frequency spark is available.

Side bend test: A soundness test in which the side of a transverse section of the weld is on the convex surface of a specified bend radius.

Silicon: Nonmetallic element used in steel making. If present in the molten pool, it will rise to the surface. Silicon may cause weld pool movement control problems and should be removed from the surface of the completed weld before making another pass.

Silicon-controlled rectifier (SCR): Solid state diodes used to change alternating current to direct current.

Single-phase current: A single sine wave of alternating current.

Slag pocket: A piece of slag (flux) of indefinite size trapped in the cooling weld metal.

Slope: Slant of the volt-amperage curve and operating characteristics of the power supply under load.

Slopers: Used in semiautomatic and automatic welding operations to slope welding current upward to the required current level and downward from the required current level.

Solenoids: Electrically operated valves used to start and stop the flow of shielding gas from the gas supply to the welding gun.

Solid state: Use of solid components for generation of power and control of sequential circuits.

Soluble dam: Paper dam used to plug tubes and pipes for purging. Flushing the tubes after welding dissolves the dam.

Solution heat-treated: Material that is heated to a predetermined temperature for a suitable length of time to allow some element in the material to enter into "solid solution." The alloy is quickly cooled to hold the element in the solution.

Spatter: Small pieces of metal that are ejected from the molten pool and attach to the base material.

Spray arc mode: Method of weld metal deposition as small droplets of metal are surrounded by slag and smoke.

Stainless steels: Family of iron alloys that resist almost all forms of rusting and corrosion.

Steel: Alloy of iron and carbon with varying amounts of other alloying elements for specific mechanical properties.

Stepover distance: Measurement of weld bead placement when surfacing a material to obtain a relatively flat crown bead.

Stickout: Length of the unmelted electrode extending beyond the contact tip in the gun. Also called electrode extension.

Straight polarity: Arrangement of direct current arc welding leads so the work is positive and the electrode is negative. *See* direct current electrode negative.

Strain-hardened material: Material strained by stretching, pulling, or forming to produce a grain structure with higher mechanical properties.

Strength-to-weight ratio: Material strength in relation to material weight.

Stress relieve: A heating operation that requires heating the weldment to a specific temperature for a period of time to relieve the stresses formed during the welding operation.

Stringer bead pattern: See stringer beads.

Stringer beads: Weld beads made without oscillation (side-to-side motion).

Suckback: Concave root surface in a full-penetration groove weld. Also called concave root surface.

Surfacing: Applying material to the surface of another material for protection from chemicals, heat, wear, and rust.

Swaging: Changing the shape of a material with mechanical tools such as hammers or dies.

T

Tachometer: Device used in welding to determine speed in inches per minute. Defines travel speed of the gun, part, or filler material.

Tack weld: Weld made to hold parts of a weldment in alignment until the final weld is made.

Taps: Male and female connection devices designed to complete a welding circuit for various amounts of welding current.

Teflon™: Insulating material used to make gaskets and insulators for welding guns. Teflon tape is also used on threaded inert gas connections to make an effective seal.

Tempering beads: Stringer-bead welding passes made on the crown of a completed weld to reduce the grain size of the crown bead. It is often ground off the part later. Tempering reduces crown bead tensile strength and improves ductility.

Tempers: Ferrous materials with various mechanical properties made by a series of basic treatments. Usually materials with decreased hardness and increased toughness.

Tensile test: Determines the mechanical properties of metal. The test is done by placing a specially designed piece of metal in a tension machine and applying a load until the part fails.

Terminal blocks: Blocks of copper or bronze used for attaching welding power and ground cables.

Thermal-cut material: Material cut by a melting process. Include oxyfuel gas, plasma-arc, or carbon-arc processes.

Thermal overload protector: Heat-sensing device that opens the welding circuit on a welding machine when a certain temperature is reached. It is used to protect the machine from damage caused by excessive heat.

Thermal treatment: Treatment of metal using heat.

Three-phase current: Type of alternating current used to operate welding power supplies.

Timers: Mechanical or electronic devices used to sequence the functions of various welding machines and other equipment.

Titanium: Light, silvery, lustrous, very hard, corrosion-resistant metallic element.

Tolerance: Permissible variation of a characteristic, parameter, or variable.

Toluene: Colorless, water-soluble, flammable liquid that has a benzene-like odor. Used for removing oils, grease, and paint from material.

Tractor: Electrical-mechanical device used to transport welding equipment along a weld joint.

Transformer: Device used in welding power supplies and equipment to change both voltage and current from one level to another.

Transverse crack: Crack that extends across a weld joint or the parent metal.

Triangular weave bead pattern: A welding technique used when welding vertical uphill fillet welds on thick material.

Trichloroethylene: Colorless, poisonous liquid generally used in a vapor state to remove oils, grease, and paint from material.

Tube sheets: End plates of tube bundle assembly.

Tungsten: Rare metallic element with a very high melting point.

Turbulence: Disturbance in the smooth flow of shielding gas, as well as irregular changes in speed and direction of flow that allow atmosphere to enter the weld zone.

U

Ultrasonic inspection: Nondestructive testing method that uses high-frequency sound waves to determine the interior quality of a weld.

Underbead cracking: Cracking in the parent metal adjacent to the weld, usually caused by the absorption of hydrogen into the weld and parent metal during welding.

Undercut: Groove melted into a base metal next to the toe or root of the weld and left unfilled by the weld metal.

V

Variables: Factors that affect welding sequence and weld quality.

Variable voltage power supply: Varying the welding voltage (arc gap) changes the output current level of a constant current power supply.

Visible stickout: Distance between the end of the gas nozzle and the end of the electrode. It is the portion of the electrode viewed by the welder using a gun with a recessed contact tip.

Visual inspection: Inspection of a material or weld through visual observation.

Voids: Holes or areas within the completed weld.

Volt-ampere curve: Operating characteristics of a welding power supply established by the manufacturer's design.

Voltage-sensing: Refers to a wire feeder used with a constant current or constant voltage-type power supply.

Voltmeter: Instrument that registers the amount of arc gap or distance between the electrode tip and workpiece. Low voltage indicates a close gap, and a high voltage indicates a longer gap.

Volts alternating current (V ac): Designated on welding power supply data sheets, labels, and operating manuals to identify the amount of primary voltage required to connect the machine to the utility power.

W

Warranty: Period of time a manufacturer guarantees a product.

Wash beads: Beads made with oscillated (side-to-side) movements to widen the beads. Also called *oscillated beads*.

Water-cooled gun: Uses water to cool the welding gun head. These guns operate at a higher duty cycle than air- or gas-cooled guns.

Weave bead pattern: *See* wash beads.

Weld chemistry: Final chemical content of the weld.

Weld cycle: Complete series of events involved in making a weld.

Weld joint: Junction of members or edges of mating parts to be joined.

Weld pool: Molten part of the weld during the welding sequence.

Weld reinforcement: Material applied in excess of the quantity required to fill the weld joint.

Weld root: Deepest part of the weld into the parent metal.

Weld schedule: A form used to record welding parameters, welding variables, and other data so a particular weld can be duplicated at a future date.

Weld shaver: Rotary tool designed to remove weld crowns from a groove weld.

Weld symbol: Part of the welding symbol that shows what type of weld is to be placed in the joint.

Weld throat: The distance from the top of the weld (crown) to the bottom of the weld (root).

Weld zone: Area immediately adjacent to the weld joint.

Welder: Person who performs a manual or semiautomatic welding operation.

Welder qualification: Based on the ability of the welder to utilize a proven welding procedure using a specified process and the proper techniques to complete a satisfactory weld.

Welding arc voltage: Actual voltage across the arc during welding.

Welding gun: Trigger or switch-operated mechanism held by an operator, or mounted in a machine, that transfers electrical current to an electrode and may contain a gas nozzle to direct shielding gas around the molten pool.

Welding operator: Person who operates automatic welding equipment.

Welding procedure: The detailed methods and practices, including all welding procedure specifications, involved in the production of a weldment.

Welding Procedure Specification (WPS): A document providing the required welding variables for a specific application to ensure duplication by properly trained welders and welding operators.

Welding schedule: A written statement, usually in tabular form, specifying the actual welding parameters to be used and the welding sequence for performing a welding operation.

Welding speed: Rate at which an electrode passes along a joint.

Welding symbol: Standard graphic representation placed on a welding drawing to indicate type of weld joint, placement of weld, and type of weld to be made.

Wire feeder: Electromechanical device designed to feed spooled filler material into a weld pool at a controlled rate of speed.

Wrought material: Material made by processes other than casting.

Y

Y-valve system: Used in single welding stations to change gases that are different from standard mixes.

Index